ALL ABOUT OPTIONS

The Easy Way to Get Started

THOMAS A. McCAFFERTY

Second Edition

McGraw-Hill

New York   San Francisco   Washington, D.C.   Auckland   Bogotá
Caracas   Lisbon   London   Madrid   Mexico City   Milan
Montreal   New Delhi   San Juan   Singapore
Sydney   Tokyo   Toronto

Library of Congress Cataloging-in-Publication Data

McCafferty, Thomas A.
　　All about options / Thomas A. McCafferty. — 2nd ed.
　　　　p.　cm.
　　Rev. ed. of : All about options / Russell R. Wasendorf & Thomas A. McCafferty
　　Includes bibliographical references.
　　ISBN 0-07-045543-0
　　1. Options (Finance)—United States—Handbooks, manuals, etc.
　I. Wasendorf, Russell R.　All about options.　II. Title.
　HG6024.U6W37　1998
　332.63'228—dc21　　　　　　　　　　　　　　　　　98-9546
　　　　　　　　　　　　　　　　　　　　　　　　　　　CIP

McGraw-Hill
A Division of The *McGraw·Hill* Companies

　7　8　9　0　　DOC/DOC　　0　3　2　1　0

ISBN 0-07-045543-0

The sponsoring editor for this book was Stephen Isaacs, the editing supervisor was Donna Namorato, and the production supervisor was Suzanne W. B. Rapcavage. It was typeset in 11/13 Palatino by Hendrickson Creative Communications.

Printed and bound by R. R. Donnelley & Sons Company.

This publication is designed to provide accurate and authoritative information in regard to the subject matter covered. It is sold with the understanding that the author nor the publisher is engaged in rendering legal, accounting, futures/securities trading, or other professional service. If legal advice or other expert assistance is required, the services of a competent professional person should be sought.

　　　　　　　　—From a Declaration of Principles jointly adopted by a Committee of
　　　　　　　　the American Bar Assocation and a Committee of Publishers.

McGraw-Hill books are available at special quantity discounts to use as premiums and sales promotions, or for use in corporate training programs. For more information, please write to the Director of Special Sales, McGraw-Hill, Professional Publishing, Two Penn Plaza, New York, NY 10121-2298. Or contact your local bookstore.

 This book is printed on recycled, acid-free paper containing a minimum of 50% recycled de-inked fiber.

To Carol, Cynthia, Monica, Colleen, and Sadie

Other Titles in McGraw-Hill's "All About..." Series

CONTENTS

First, we need to address the following questions—

- ◆ What Are Options?
- ◆ Do They Have Economic Value?
- ◆ How Should an Investor Use Stock Options and Options-on-Futures?
- ◆ What Are the Pros and Cons, Risks and Rewards?
- ◆ Is Option Trading Gambling?

Options, in general, are the legal rights, acquired for a consideration, to buy or sell something at a predetermined price by a certain time in the future. Options are very common in real estate, for example. Developers will take options on certain pieces of property as they plan projects. It gives them time to obtain zoning approvals and capital for development. The option locks in the price and buys time.

Real estate and most other options are negotiated from scratch by the two, or more, parties involved. The buyer and seller need to agree on all the terms or specifications, not to mention price. Reaching an agreement may take anywhere from a few hours to several months. Negotiations can break down, resulting in no agreement or option.

There is another type of option. It is called an exchange-traded option and is the province of this book. This type of option differs from the other in that all the specifications are prearranged, except price. For stocks, options generally "cover" 100 shares of the underlying stock; options-on-futures are on the underlying futures contract. All the other specifications—delivery date, place and type of delivery, quality, etc.—are clearly defined. The only variables are the number of options to be bought or sold and the price.

Since buyers or sellers decide for themselves on quantity, the only thing left to negotiate is price. It is this characteristic of uniformity that makes the stock and futures exchanges possible. The exchanges, in turn, render society in general, and the business world in particular, a very important economic service—the discovery of price. Trading in options and the underlying stocks or futures contracts tells users, planners, reporters, economists, speculators, investors, government officials, and anyone else interested, the current value of a company's stock or a commodity, along with projected price trends into the future.

If you are a user or processor of soybeans, copper, money, or any one of the over 60 options-on-futures traded in the United States, you can easily check the futures market to obtain the current price and get a good insight into what prices are expected to be 3, 6, or 12 months from now. A look at any of the thousands of stock option prices tells you what savvy traders think about the prospects of individual stocks. Or you might study one of the many composite indexes of stocks, commodities, utilities, foreign currencies, etc. (or the options on these indexes), to get a feel for the price direction of an entire market.

Another possibility might be that your analysis indicates you are facing some serious financial risks. Options could provide some relief. This is the second very important function of the options markets. Coupled with risk management is the ease and convenience of buying and selling options.

Let me give you an example. You own a quantity of something. It could be an agricultural product (grains, livestock) or a financial entity (a stock portfolio, Treasury bonds). Analysis of the price trend indicates that the commodity you own could lose value over the next 6 months. You don't want to or can't sell the commodity until then. What do you do? You can offset the risk of ownership by taking a short position in an option. You freeze the price through the purchase of the proper option. This is known as hedging and will be discussed in detail in the chapter on strategies.

The important point is that the options markets can be used to pass the risk of ownership of stocks and commodities to someone, the speculator, who wants to accept the risk. The speculator's objective is to make money from the acceptance of the risk. The beauty of using exchange-traded options is the ease with which the transfer can be accomplished.

For example, you don't have to go out and find people to accept the risk. They are immediately available at the exchanges. When you're ready to close out your position, you can do it with a single phone call to your broker.

If you are dealing with a foreign currency or foreign financial instrument, you do not have to worry about setting up an asset custodial relationship in the foreign country. Nor is there any problem with transferring money to another corner of the world or finding an English speaking broker or currency conversion. Additionally, there are ways, which I'll discuss later in this book, for trading these markets 24 hours a day. This can become important when major economic or political events occur while U.S. exchanges are closed. The fall of the Berlin Wall and the outbreak of Desert Storm come to mind as times when this facility might be utilized.

I would be remiss in my discussion of the benefits of options if I didn't mention how they can improve the efficiency of your use of capital. Buying an option on any underlying entity can be done for a fraction of the cost of obtaining outright ownership of the entity. Keep in mind, when you buy an option, that you are only acquiring the right to assume ownership before a given date at a predetermined price. You don't own the entity.

By the same token, when you're ready to exit your option positions, it is nothing like selling an entity you actually own and possess. There are no shares to sign and deliver (stocks). Nor do you have to truck your commodity to the elevator (grains) or stockyard (livestock) or remove it from a safety deposit box and have it assayed (precious metals). You avoid all quality and quantity confrontations that can occur when commodities are handled, rehandled, stored, and shipped. It is not uncommon for farmers to sell their crops right out of the field and then buy options to replace the actual grain in order to partake of the seasonal price rises that normally occur after harvest. Gone are the worries of spoilage, cracked kernels, and rodents—not to mention the hassle of handling and rehandling the crop and trucking it to the elevator.

LIMITED RISK AND UNLIMITED REWARD

So far, I have only briefly alluded to speculators, the people who make the whole system work. Speculators assume all the risk others wish to avoid. Most of you readers will fall into this category.

What motivates speculators to buy options is that they know in advance to the penny what their financial risk will be—but their potential profit is theoretically unlimited. They risk the cost of the option, known as the premium, and the transaction costs, which are broker-age commissions and fees. The reward depends on how much value the underlying entities gain before the speculators take their profits.

A speculator could conceivably buy an option for $300 and sell it for $3,000 or $30,000 within 30 days. This is certainly possible and has happened, although it is not normally the case in the options markets. The normal course of business is for options to expire worth-less. In other words, the speculator invests $300 and loses $300. Experienced option traders usually offset their positions before the options expire worthless. The traders will invest $300 and close the position when it has lost $100 or $150. They'll continue to do this until they choose some winners. I'll talk about this in detail in the discussion on how to use the law of probability to become a net winner as an option trader.

Even the concept of unlimited profit potential needs some explanation. Theoretically, options have unlimited profit potential—the emphasis here being on the word "theoretically." But prices for stocks or commodity futures normally do not rise or fall unlimited amounts. At some price, they become either too expensive or too cheap. For example, during the Great Depression, the price of corn fell so low that corn was cheaper to burn than coal. And that's what some farm families did. On the other end of the spectrum, when corn gets more expensive than wheat, livestock producers feed their animals wheat. The demand for and the price of corn eventually return to more reasonable ranges. The same is true for stocks. If the price of a company's stock drops below the liquidation value of that company's assets, what will happen? Or what will happen if the price of a stock rises so high no one can afford it? The answer is the same in both cases: The market adjusts the price accordingly.

Additionally, there is the profit taking motive to factor into the supply-demand equation governing the price of stocks or commodities. Investors who got in early, on a bull or bear move, will take their profit at some point. Eventually, the profit takers turn the market around.

What option traders have going for them is predefined risk, rather than limited risk. On the profit side, they have the prospects of an excellent return, rather than unlimited profits. Experienced investors seek outstanding risk-to-reward ratios. If they can invest

$500 with the potential of making $5,000 in a few weeks, or a 1:10 risk-to-reward ratio, they'll consider it.

Another reason for buying options that appeals to speculators is the staying power it gives them in adverse markets. Buying options-on-futures, unlike futures themselves, does not require margin money or a good-faith deposit. The risk is fixed, as previously stated, to the premium and transaction costs. If the markets move against your position, you can hold the option to expiration without having to put more money into the investment. With futures, traders have to replace any money lost when the market moves unfavorably. If you have a very strong conviction that the market being traded will eventually be profitable, an option, if it has enough time to expiration, provides the staying power to hang on to the position.

IS OPTION TRADING GAMBLING?

If investors do not do their homework, arbitrarily buy options, and do not closely follow the markets, you might as well consider what they're doing is gambling—but wouldn't this be true for any investment?

Serious option traders use the law of probability to take advantage of the distortions in the supply-demand situations of stocks and/or commodity futures contracts that regularly occur. Experience, research, and even a little help from a good broker can put the statistical odds in your favor.

Trading options on stocks and/or futures is complex. Serious investors must invest in books, newsletters, seminars, software and other educational materials. Much time is required to master option trading—but the potential rewards can make the effort a joy.

On the other hand, I have no intention of misleading you. Option trading, like trading in any other investment, entails substantial risk. Not every investor is suited for it because it is such a highly speculative venture.

Reading this book is a good start toward understanding options, but please consider it as only the beginning. When you're finished, you should have a solid overview of the subject and a good idea of where you need to turn next in your quest for success in the options markets.

CHAPTER 1

Understanding the Basics, Particularly the Value and Risks Associated with Stock Options and Options-on-Futures Trading

"Now, on the St. Louis team we have Who's on first, What's on second, I Don't Know's on third."

Bud Abbott
from Naughty Nineties *(1945 film)*

Key Concepts

* Terminology Can Be Confusing
* What Is an Option? Puts and Calls?
* Differences and Similarities between Stock Options and Options-on-Futures Contracts
* Risk versus Rewards of Option Trading

One of the funniest comedy routines of all-time may be the exchange between Bud Abbott and Lou Costello about the names of the players on the St. Louis baseball team in their film *Naughty Nineties*. The name of the first baseman was "Who," the second baseman "What," and the third bagger "Don't Know." When Costello asked Abbott "Who's on second?" Abbott responded by saying: "No, he's on first. What's on second." When asked "Who's on third?" the retort of course was "Who's on first. I Don't Know's on third." These two comedians go round and round in the hilarious confusion over the players' names. Options trading terminology can be equally confusing, but not nearly as funny.

Unfortunately, options terminology is like a grand English maze. In the beginning, it appears to be very simple and straightforward. The deeper and deeper you get into it, the more complex and confusing it becomes. Eventually you can become lost in the language, like our two friends, Lou and Bud. The danger is you might be turned upside down in the process, so every penny in your pockets falls out. The purpose of this chapter is to prevent that from happening to you.

An option is only an option. That simply means that an option to buy a piece of property has much in common with an option to buy an artwork in progress or 100 shares of a stock or 5,000 bushels of corn. The option gives you the right to buy something. Whoever owns or controls the entity being optioned wants control back at some point. Therefore, all options have expiration dates. Giving you an option prevents the person granting, or selling, you the option the right to sell it to someone else. For this reason, the grantor asks for compensation. Last of all, there is a price set at which you can buy the entity under option, if you wish to exercise your option.

In most situations we run into in daily life, options only involve buying. We take an option on a piece of property we are considering buying. We don't normally take an option on a piece of land we don't own or plan on selling. There are land speculators who take an option on a farm or house with the hope of reselling it before their option expires. But they really aren't selling something they don't own. In these cases, they control the entity or lock it out of the market until they resell it.

As you will learn as you read on, you can use options to sell what you don't own or don't plan on buying. You can buy a call or a put. The call is most like the conventional option. When you buy a call you have the right—but not the obligation—to buy 100 shares of a stock or a long position in one of the futures markets at a set price before a set deadline.

A put, on the other hand, gives you the right to sell 100 shares of a stock or a futures contract at a given price before a given deadline. Whether it is 100 shares of stock or a futures contract, a lot of people have a problem with the concept of selling something they have never owned.

The solution to this quandary is the other party to your transaction. Every option has a grantor, someone that sells you that option. The grantor is guaranteeing your ability to perform or exercise your option. If you do exercise the option you bought, the grantor must sell you the 100 shares of stock or the futures contract you bought. In some cases the grantor may actually own the stock or the futures contract. This is called a covered option. The grantor sold you a covered option. It's covered from his or her point of view, not yours. If the grantor does not own the underlying stock or futures contract, he or she must venture into the marketplace—the stock or futures exchange—and buy it so it can be delivered to you.

It is very important to get these basic ideas and keep them straight from the very beginning. Let's go over a few examples. When you initially open a position, you buy a call or a put. A call gives you the right, but not the obligation, to take a long position in the underlying stock or futures contract at a specific price, called the striking or strike price. When you buy a put, you have the same rights you have with a call, except you can assume a short position in the underlying market.

This is where the confusion often begins. One alternative the trader has is to offset or exit his or her put or call position by selling an exactly equal, but opposite, position. To offset a call, one sells a call for the exact same strike price and expiration date on the same exchange for the same stock or futures contract. To offset a put you purchased, you sell a put. All the contract specifications—underlying entity, expiration date, exercise or strike price, type (meaning put or call)—must match. One position offsets the other. For example, you think the price of corn is going to increase substantially over the next 60 days. Since your analysis indicates a price rise, you want to "own" the corn now and sell it 60 days from now at a higher price. To do this, you would go long the futures market. Rather than do that, you opt to take an option on a long on futures contract. That would be a call. Therefore, you buy one corn call option. You now have the right, but not the obligation, to assume a long position in the corn futures market at a specific price, called the strike or exercise price.

Let's get more specific. The Chicago Board of Trade (CBOT) futures contract for corn, which expires in July, is trading at $3 per bushel. Since the contract calls for the delivery of 5,000 bushels of Number 2 corn in July, it is currently worth $15,000. Your analysis indicates it will increase to $4 a bushel or $20,000 before its delivery date in July. That's a per-contract profit of $5,000 before paying brokerage commissions and exchange fees, which are commonly referred to as transaction costs. They can range between $25 and $125 per contract for futures.

Instead of buying a futures contract, you decide to buy an option on one CBOT corn option. The strike or exercise price you select is $3.10 per bushel, or 10 cents above the current underlying corn market. By being 10 cents above the market on a call, the option is cheaper than an option closer to the current trading price of the underlying entity. Let's call it a 1-cent premium, or $500 (5,000 bu. × $0.01). Also, you want to see the bull move (price advance) begin before you get too committed to a futures position.

The $3.10 corn call is considered a near-in-the-money option, which means it is near the current price of the underlying contract. If you have chosen the $3.00 option, which is the current price of the underlying contract, it would be considered to be at-the-money. The cost of an at-the-money option would be higher; let's call it 2 cents per bushel, or $1,000 for this $3.00 corn call. A $2.90 corn call would be 10 cents in-the-money.

Now that you have bought a $3.10 July corn call, what do you do with it? You have three alternatives. First, you can do nothing and let it expire worthless. Second, you can exercise and assume a position in the underlying futures market. In this case, you would be long one July futures contract at $3.10, which is the strike or exercise price.

Last of all, you could offset the position. This is where the terminology can become confusing. If you wished to offset your position, you could sell one July corn call. The term "sell" also means to write or grant an option. The writer or seller of an option is the trader who agrees to provide a futures contract to an option holder whenever that person decides to exercise the option. In the case of stocks, the writer provides the shares to the option holder when and if the option is exercised. For assuming this responsibility and risk, the option writer of the corn call receives the $500 premium. The risk is that the price of the underlying entity will increase. Therefore, a writer will have to deliver a contract that has gained more value than the premium received.

When you tell your broker you want to sell an option, are you offsetting an option position you already own or are you offering to write (as in underwrite) an option someone else wishes to buy?

All this can become even more confusing, since options are traded like futures. With futures, you can short the market. This means if you forecast that the price of a certain commodity is headed lower, you sell at a given price and buy back (offset) that position at a lower price. The difference is your profit.

For example, in our corn scenario, let's say you thought corn was headed down $1, rather than up. Using the futures market, you would short the market at $3 and go long, offsetting your short position, at $2. That equates to a $1-per-bushel, or $5,000, profit.

You could do the same thing with options. In this case, you buy a put giving you the right, but not the obligation, to take a short position in the underlying futures contract. If your projection of a $1 fall in the corn market was correct, you could exercise your option and become short in the futures market. Or you could offset your options position by selling a put that was exactly the same as the one you bought.

My point is that you must make it very clear to your broker what action you wish to take or you'll end up like Abbott and Costello. But the humor may be missing if it costs you money. When in doubt about a term, refer to the Glossary at the back of this book. Don't sell a put when you meant to write a call! See Figure 1–1.

There are 20 Magic Words to memorize:

Long call—Right to buy

Long put—Right to sell

Short call—Obligation to sell

Short put—Obligation to buy

In the above example, if you sell a put you don't own by accident, you would not be offsetting the position you have in your

F I G U R E   1–1

Comparison of Buyers to Sellers

Option Buyer versus Option Seller

Obtains the Right, but not the Obligation, to Assume a Position in the Underlying Entity	Investment Risk Equal to the Maximum Move the Underlying Entity can Muster
Investment Risk Equal to 100% of Premium Paid for Option Plus Brokerage Commission and Fees	Assumes the Risk of Delivering a Position in the Underlying Entity
Pays Premium	Receives Premium
Pays Commissions and Fees	Pays Commissions and Fees

account. You would be taking on the obligation to buy a stock or futures contract at a strike price. If that stock price jumps $10 per share, you would be under the obligation to sell 100 shares to whoever bought the options. That would be a $1,000 loss; and the option owned, which you wanted to offset, would probably be worthless, since it was a put and the stock gained $10 in value.

You may be thinking that the example above isn't realistic. It is. I've seen it happen. Moves happen fast in some markets. Investors try to react fast. Mistakes happen. A person is holding a June option long and sells a July. Or the person means to say sell and says buy in the heat of the action.

For this very reason, many brokers tape-record conversations between themselves and their clients—at least when they are taking orders. It is not a bad idea to tape your own conversations with your broker. All it takes is about $30 worth of equipment from Radio Shack, or an answering machine that has a record function. Also, be sure to have your broker repeat any buying or selling instructions you give.

While the discussion is still on terminology, here are a few more terms you should know. The value or cost of an option is composed of intrinsic value and time value. Intrinsic value is what an option would be worth if it were exercised. For example, your $3.10 corn option would have 50-cents-per-bushel ($2,500 per CBOT contract) intrinsic value when corn is trading at $3.60. It would be said to be 50 cents in-the-money. When corn moved from $3.00 to $3.10, your $3.10 option would be at-the-money. While corn is at $3.00, your $3.10 corn call is out-of-the-money by a dime. There are also deep-out-of-the-money options, which are quite a ways away from the current price of the underlying contract.

When you buy a $3.10 corn call, corn is at $3.00. You might pay $500 for it. There is no intrinsic value. The $500 is time value. The market is saying that the 4 to 6 weeks left on the option has a value of $500. For this time value to be recovered, corn must go from $3.00 to $3.20 to generate $500 intrinsic value, or 10 cents per 5,000 bushels over the $3.10 strike price. At that point, depending on what traders think the market is going to do and how much time is left before the option expires, some additional time value would be added to the $500 intrinsic value.

The intrinsic value of a put is figured the same way, only from the opposite side of the market. A put has intrinsic value when its strike price is above the current price of the underlying contract. For example, a $3.10 corn put would have a dime or $500 (5,000 bu. × 0.10) intrinsic value when corn is at $3.00 per bushel. It may have some time value as well. The longer the time left on an option, from as long as a year to as little as a day, the higher the time value usually is. Or the more time left on an option, the greater the chances are of the option ending up in-the-money. At least, that is the conventional wisdom.

This concept does not quite fit deep-out-of-the-money options. Each exchange has its own definition of what this type of option is, but the most common definition is an option that is two strike prices out-of-the-money plus one more strike price for each month left to expiration. For example, corn call strike prices are a dime apart. If the current futures price is $3.00 and the strike prices for calls are $2.80, $2.90, $3.00, $3.10, and $3.20, the $3.20 would be classified as deep-out-of-the-money for the nearby month. The time value—their only value—is usually low, unless the option has a decent amount of time left and there is a lot of price activity occurring.

The reason for this is twofold. First, it takes most markets awhile to move to a point where these distant strike prices begin to become attractive to speculators. Second, once the underlying market starts to get within range, the time value builds quickly.

With all this talk about intrinsic and extrinsic (time) value, you may be wondering how option prices are set. Options are traded at various exchanges throughout the world—stock options at stock exchanges and options-on-commodities at futures exchanges, plus options have their own exchange, the Chicago Board of Options Exchange (CBOE). Additionally, there are regional exchanges to handle options on regional stocks. Some of the options of the financial futures contracts are traded on the futures markets, even options on stock indexes, while some options on foreign currencies and stock indexes are traded at stock exchanges. You'll find a review of the exchanges and the option products they offer in Appendix 3. Also keep in mind that new option contracts are always being added and unpopular ones dropped. It's impossible to maintain a definitive listing.

The primary price discovery process utilized by exchanges is known as "open outcry." This is where floor traders in the trading pits (or rings) call out their bid or ask prices to all other floor traders in the pit. When a match is made, a trade is consummated. The "outcry" of orders has its origins in the early rules of the exchanges, going back to the eighteenth century in the United States. The rules required that all orders be shouted out so that any member would have the opportunity of filling the order. As the exchanges grew and became busier, a system of hand signals was developed to augment the oral tradition. The floor brokers record all their buys and sells, referencing the organizational badge of the person on the other side of the trade. All these "deals" are fed into the computer of the clearing firm which balances each day's trading.

Spotters, who work for the exchanges and are located on the perimeter of each pit, transmit via radio the stream of prices generated by the trading. This is the flow of prices you see on cable television or price quotation systems. The futures exchanges capture the bid and ask prices, while stock exchanges only transmit fills or consummated sales. With futures prices, you will occasionally see on a screen the price you expected to get only to find out later your order was not filled. This occurs because of the procedure of transmitting both bid and ask prices. The extreme high volume of the futures markets makes this a necessity.

You, as an option trader, call your order to your broker. Your broker relays the order to an order desk at the appropriate exchange. From here it is signaled or a runner takes it to a floor trader in the proper pit. Once the floor trader fills it, the process is reversed. In high-volume markets, your order could conceivably be completed and reported back to you in minutes. Slower markets (lower trading volume), particularly for strike prices out-of-the-money or options in the distant months, take longer. There are even occasions when an order cannot be filled.

The exchanges are continuously experimenting with technology to improve the speed and accuracy of options trading. For example, the futures exchanges are developing electronic order recorders to replace handwritten records. Another example would be GLOBEX, which might be characterized as a secondary price discovery process.

GLOBEX is an international electronic system for futures and options trading which opens when the regular United States markets close. It is a joint venture of the Chicago Mercantile Exchange (CME) and the Chicago Board of Trade. The Marche à Terme International de France (MATIF), the French commodity exchange, also participates. The big four New York commodity exchanges, i.e., the Coffee, Sugar and Cocoa Exchange; the Commodity Exchange; the New

York Cotton Exchange, and the New York Futures Exchange, have agreed in principle to become active. Additionally, several other European and Asian exchanges are very interested. Another major player in GLOBEX is Reuters Limited.

GLOBEX provides the hardware, software, and computer know-how and operates the system. The control center is in Chicago. The most obvious difference between GLOBEX and the regular exchanges is that orders are matched electronically, rather than via open outcry. It matches orders based first on price and second on time of entry, with orders at the same price being matched so that the order received first at the host computer is matched first.

Another big difference is that it is multinational and multiexchange. The laws controlling the trading activity are subject to the laws of the country of the exchange on which the option you are trading is listed. Therefore, if you're trading the 7-year ECU interest rate option contract on the MATIF, your trade falls under the laws of France. Additionally, the rules and regulations of the listing exchange govern your rights and the trading rules.

On the stock side, most orders are entered electronically. The CBOE uses the Order Routing System (ORS) to collect, store, route, and execute public customers', which means non-broker dealer, orders. The ORS automatically routes option market and limit orders of 10 or fewer contracts to the Retail Automatic Execution System. These orders receive instantaneous execution at the prevailing market quote and are confirmed almost immediately to your brokerage firm. Larger orders are routed to floor brokers via your brokerage firm's floor order desk and to the order book official's electronic order book. If within market range, orders are promptly filled.

It is my opinion that more and more trading has to be done electronically in the years ahead. Some brokerage firms, for example, accept customer orders electronically already. And the Internet is just beginning to show its promise, as security and reliability improve daily.

STOCK VERSUS FUTURES OPTIONS

There are some important similarities and differences between stock options and options-on-futures contracts. First, both represent a standardized contract. For stocks, it is usually 100 shares of the underlying stock. An option on Exxon, for example, would be an option on 100 shares of that stock. With futures, the option is on one futures contract for that commodity. Any 90-day Treasury bill option on the International Monetary Exchange (IMM) contains $1,000,000 in T-bills. All the specifications—namely, quantity, quality, commodity, delivery place, and delivery date—are uniform. See the list of futures option specifications in Appendix 4. All that the traders (the buyers and sellers) need to agree on is price. It is this concept that gives all the exchanges the liquidity needed to be useful.

Another important concept is "settlement," which is the daily procedures for resolving accounts between members. You, as a customer of a brokerage firm, are represented at the exchanges by a member. Therefore, any trades you make must be settled at the end of the day. Once settled, which means your buy or sell tickets are matched with someone else's tickets, the clearing firm guarantees your position. The practice of making the clearing firm one side (buyer or seller) of every trade is what makes the exchanges so liquid. It relieves you of the responsibility of finding a buyer for your sell order and vice versa.

There are two basic methods of figuring account settlements: stock type and futures type. A stock-type settlement requires that the entity being purchased is completely resolved before profit or loss is experienced by the owner, or before gains can be withdrawn from your account. For example, you purchase 100 shares of John Deere stock at $25 per share, or $2,500. The stock goes up to $30, or an increase of $500. Later it retraces to $20 a share. Eventually, you sell it for $25. You broke even, notwithstanding commission and fees.

When the stock gained $5, you could not take your profit, nor were you asked to make up the $5 loss when it sunk to $20, unless you were trading in a margin account and the $5 loss per share signaled a margin call.

The futures-type accounting is somewhat different, and you should be familiar with it in case you ever exercise one of your futures options. Futures accounts are totaled by what is known as daily mark-to-the-market. For example, you are long (it "be-longs" to you) one Chicago Board of Trade silver contract at $5 per ounce for 5,000 ounces, or a total contract value of $25,000. To hold this contract, you have deposited $2,000 margin money. Margin is a goodwill deposit, not to be confused with a down payment. Then silver increases to $6 per ounce. Your account is credited $5,000, which can immediately be withdrawn if you wish. Or if silver decreased $5 per ounce, which would be a loss to you of $5,000, you might have to add money depending on the balance in your account.

Let's say the required margin was $2,000 for the silver contract and that is also the balance of your account. If the silver contract lost $5,000, it would wipe out the $2,000 in your account, plus your account would be on debit for an additional $3,000. If you wanted to continue to hold the silver position, you'd have to make up the $3,000 and add $2,000 for margin. You would only be given 24 hours to produce the funds or your account could be closed out and you'd be responsible for the debit.

Longs and shorts in futures options are treated differently when it comes to margin. If you are a seller (writer or grantor) of futures options, your account is governed by the futures-type settlement procedure. The sellers of futures options underwrite (or stand behind) the buyers of options. To ensure performance, the exchanges require option sellers to maintain enough margin money in their accounts to be able to immediately accept the future position(s) they have sold. When you grant options (sell or write an option), you are told you can be assigned a futures position at any time. If you write calls and the buyer exercises the option, the buyer is assigned a long position and you, the seller, are assigned the short or opposite side of the trade. Sellers in the futures must have enough margin money in their accounts to accept the assigned futures contract, but they can withdraw any excess. When buyers of puts exercise their options, sellers of these traders are assigned long positions at the strike price of the option.

If you are long futures options, you cannot withdraw unrealized gains, as you can if you are trading the actual contracts or as the seller can.

If you short stock options, the rules get complicated and the margin requirement varies by type of option. For example:

Type of Option	Margin
Covered option	Zero, no margin required
Uncovered equity option	Premium + 20% of current market Option value (CMV) – any amount option out-of-the-money
Uncovered nonequity option	Premium + x% of CMV or face value – any amount option out-of-the-money

Nonequity options can be indexes, foreign currencies, or Treasury bills, notes, or bonds. With Treasuries, you use face value to calculate margin; with the others you use CMV, and the percentages range from 10 percent to $\frac{1}{20}$ percent of face value.

You will probably never be required to calculate a margin, but at times you may wonder how it was arrived at. Futures margins are equally mysterious. They are set by a margin committee at each exchange after reviewing a "risk array" generated by a computer program known as SPAN. Always keep in mind that margin requirements can and do change without notice and the rules stated above can change just as easily. The reason for change is to protect the integrity of the markets based on the amount of risk (volatility), real or perceived. Also there is a difference between initial and maintenance margins; the latter is usually a little lower.

The term "settlement" also refers to the final, published price of the day for each stock or commodity contract traded. A committee at each exchange sets the settlement price. When trading volume is high, there can be more than one closing price. This occurs most often in the larger trading pits, when a flurry of orders enters the pit just before the bell rings ending trading. A group of traders in one corner may agree on a price different from one agreed upon by another group at the other side of the pit. These prices are usually very close, within pennies in most cases. The price settlement committee decides on which price is most representative of the day's closing activity.

Now option traders, as a rule, do not exercise their positions unless they are in-the-money, or at least at-the-money. This means the positions have intrinsic value. To the sellers (writers or grantors), it means someone will be assigned an equal, but opposite, position that is losing money—that person is immediately at financial risk. He or she must stand behind these losing trade(s). Option writers usually react by offsetting these positions as soon as possible. I'll discuss this procedure more in the strategy chapters later in the book.

You may be wondering about the premium the sellers receive. When sellers grant options, the buyers pay the premiums to the sellers, via their brokerage firm. The brokerage firms, which are handling the trades, credit the seller's trading account for the premiums and debit it for the margin that underwrites the options, plus transaction costs (brokerage commissions and exchange fees).

Here's a simple example of a futures-type trade. Trader Jones's analysis indicates soybeans will gain $2 per bushel over the next 90 days. Soybeans are currently trading at $5 per bushel. He buys a call at-the-money for $1,000. The $2 projected gain on a 5,000-bushel contract is $10,000. He is risking $1,000 to make $10,000, a 1:10 risk-to-reward ratio.

Trader Smith doesn't see it that way. She believes soybeans will trade in a sideways pattern for the next 60 days. She decides to sell or underwrite Trader Jones's call. Her account is credited the $1,000 premium and debited $850 in margin money, a brokerage commission of $45, and $5 in fees. She has $100 excess capital in her account which she withdraws, leaving $900 in her trading account.

The following table shows you what happens to the option's intrinsic value and the seller's account.

Prices	Seller's Account	Intrinsic Value of Option
4.50	900	0
4.75	900	0
5.00	900	0
5.20	−100	1,000 (Breakeven point)
5.25	−350	1,250
5.50	−1,600	2,500
5.75	−2,850	3,750
6.00	−4,100	5,000

If soybean prices go lower, the seller's premium is protected, but it doesn't gain in value. The maximum profit the seller can expect is the premium.

The buyer has unlimited profit potential, assuming the price of a commodity is unlimited, which of course is an unrealistic assumption. When prices become too high, substitutes are used or users reduce or halt processing the commodity. Nevertheless, soybeans have traded over $10 per bushel.

In the above example, the buyer of the option has two alternatives. Once the option goes into-the-money, Trader Jones could exercise his option and assume a long position in the soybean futures market. At that time, he would have to put up margin money. His second option would be to offset his call. Besides the intrinsic value, there would be some time value, depending on how much time was left before the option expired and the volatility of the market. To offset the call, Trader Jones sells a call (not to be confused with writing or granting a call) of the same strike price, same expiration month, and same exchange—an equal, but opposite, position. As mentioned at the beginning of this chapter, terminology can be confusing. You offset a call you bought by selling a call. You offset a put you bought by selling a put. You don't offset a call by buying a put. Nor do you offset a put by buying a call. You can short puts or calls and buy them back to offset.

Buying Options:

Position	Offsetting Position
Long call (bought a call)	Short call (sold a call)
Short call (sold a call)	Long call (bought a call)
Long put (bought a put)	Short put (sold a put)
Short put (sold a put)	Long put (bought a put)

Selling Options:

Position	Offsetting Position
Sold (wrote, granted) a call	Short position in futures or stock market
Sold (wrote, granted) a put	Long position in futures or stock market

There are good reasons, or more correctly strategies, for being either a buyer or writer of options, which I'll discuss later. For now, it's time to finish the discussion on the differences and similarities between stock options and options-on-futures.

One of the big differences between stock and futures options has to do with the agents who are licensed to broker these two different investments. Brokers usually specialize in either stocks or commodities. There are some who are legally registered to broker both, but they are rare and usually prefer either stocks or futures. The burden of staying current with the enormous amount of information that could influence the direction of all these markets is staggering.

Additionally, the futures and the stock markets are considered by most professionals to be negatively correlated. This means they normally move in opposite directions. When the stock indexes move north, the commodity indexes head south. Being countercyclical to each other increases the difficulty of trading both simultaneously because the timing of the trend changes is not exact.

There is some middle ground, like the stock indexes. They are a compilation of groups of stocks. Some are traded on futures markets, while others are traded on stock exchanges. Some stock traders use them to hedge large stock portfolios, a concept I'll detail when I talk about strategies.

There is another very distinct difference between the stock and commodities markets that illustrates just how useful and versatile the options markets really are. There is no futures market for stocks, just options. Some professional stock option traders can't understand why commodities have a futures market, now that options-on-futures can be traded. These savvy traders expected the options-on-futures market to totally engulf the futures markets.

Think about it for a moment. Stock options are options on the "cash" or "spot" market for stocks. You buy an option directly on the stock. The option gives you the right to assume a position in the stock market for x number of shares.

This line of thought provokes these professional traders to ask why not just buy and sell options on cash commodities. The futures market to them seems redundant. An option-on-futures is an option on a contract that is a contract on the cash commodity. Additionally, they reason, options are just as versatile as the futures market. They may be right in their assessment in the long run, but for now there is a strong futures market in commodities that you need to become familiar with in order to trade its options.

The markets for options in stocks, futures, and indexes are so complex that they are beyond the scope of most individuals from an analytical and research point of view. For example, the futures markets include the grain, meat, metal, petroleum, food, fiber, debt instrument, financial, and foreign currency complexes. The thousands of stocks available can be grouped by such major categories as pharmaceuticals, automobiles, steel, retail, manufacturing, etc. I believe you must understand and track the markets you trade to be successful. Therefore, you must narrow your scope and develop a relationship with a broker with similar interest and experience.

Option trading is a specialty within itself. You'll learn, as you study this book, that you shouldn't expect just any broker to supply the special expertise option trading requires. I've seen brokers who refused to trade options because they lacked a thorough understanding of them.

Another one of the differences between stock and futures options illustrates the above point. Stock options are basically the same from one stock exchange to the next. Futures options can vary between exchanges. Basically, I'm talking about changes in the specifications of the underlying futures contract. For example, the MidAmerican Exchange trades contracts that are approximately half the size of the ones on the other futures exchanges. This may be an exchange to consider when you first start out because of the small size of the contracts—less size, less risk.

Now let's get back to some important terminology you should know to understand the material in the chapters ahead. What if you walked in on the following discussion of strategy between your broker and his manager, as they reviewed your options positions:

"Don't you think you're too long delta? Didn't you want to be delta neutral?"

"I'm just trying to maintain a long gamma. That's what has caused the long delta."

"Your problem is you're trading on price, not volatility, particularly with those out-of-the-money options."

"Okay, okay, I'll get delta neutral at the open. I just hope it doesn't mess up the kappa!"

Does any of this remind you of Abbott and Costello's conversation? An option's delta factor provides an insight into the relationship between the option and its underlying entity. The delta measures how fast the price of an option changes compared with its underlying stock or futures contract. If a stock price increases or decreases $10 per share or a commodity futures price moves 10 cents per unit, how much does the price of the option change?

Deltas for calls range between 0 and 1 and are usually converted to percentages. For example, a delta of 25 or 25 percent means that the price of a call will move one-fourth as much as the price of the underlying entity. In our example above, if a stock gained $10.00 a share, the option on that stock with a delta of 25 will increase $2.50. The same would be true for an options-on-futures. If the futures price increases 10 cents, the call's premium gains by 2½ cents. If prices decreased, the premium of the calls would decrease at the same speed.

Puts—options to take a short position in the underlying entity—react similarly, except they move in the opposite direction of the underlying entity. As stocks or futures contracts gain, puts move further out-of-the-money. Therefore, they lose value as the underlying entity gains. For this reason, their deltas range from 0 to –1 or 0 to –100 percent. If you were holding a put on the stock described above and it gained $10.00, the put with a –25 percent delta loses $2.50. But if the stock lost $10.00, the put increases $2.50.

In everyday trading situations, most traders don't think in terms of positive or negative delta. (The exception is when you're hedging calls against puts or vice versa, as I'll get to shortly.) They just calculate the delta and keep themselves aware of where the option is (in-, at-, or out-of-the-money). They then know a positive move in the underlying entity increases the premium on calls and decreases the premium for puts. A decrease in the value of the underlying entity does the opposite.

The delta factor is calculated by working backward from market activity. You divide the amount of price difference of your option by the amount of price difference in the underlying entity. For example, if the premium of your option increases a nickel per bushel or $5 per share, when the futures price increases a dime or the per-share price gains $10, you'd have a delta of 50 or 50 percent ($0.05 divided by 0.10 or $5 divided by $10 = 50 percent). Since the price of the underlying entity is constantly changing, so is the delta factor.

Since much of the premium value of an option depends on whether it has intrinsic value or how close it is to having intrinsic value, the delta of typical options at-the-money will be about 50 percent. In-the-money options, since they can be immediately exercised for the underlying entity, have deltas at or approaching 100 percent. Deltas never exceed 100 percent because they cannot be more valuable than the underlying entity. Options that are out-of-the-money or deep-out-of-the-money have deltas from 50 percent to zero. These options have nothing of value but time.

One small point needs clarification. When I use the term "option," I am referring to American-style options. These options can be exercised at any time prior to expiration. There are also European-style options; these can only be exercised at specific, predetermined periods of time. It is usually only one or two days prior to expiration. The buyers or holders of European-style options can't exercise them when they get in-the-money or whenever they feel like it. For example, the foreign currency options on the Philadelphia Stock Exchange are of the European style. I do not recommend European-type options for novice traders. There is a bountiful selection without them and their restrictions. Just make sure your broker knows you are not interested in any European-style options.

The delta factor is naturally influenced by everything that influences the value of an option. Specifically, this is the time to expiration, the price volatility of the underlying entity, interest rates (particularly for options on stocks), and any seasonal factors. But most of all, it is the relationship between the strike price and the current market for the underlying entity that counts.

With all this in mind, it is not surprising that many option traders use the delta factor as a screening tool when selecting which option to purchase. For example, you're considering buying a soybean call. You think between now and the middle of February bean futures will increase by $2.00 per bushel. When you calculate the delta factor of the two options you are considering, the first one is 0.50 and the second one is 0.60. This means that if your price projection of $2.00 is correct, the value of the first option will increase by $1.00 and the second by $1.20. If the second option costs only 10 cents per bushel more, it is probably a better bargain—all other factors being equal—than the less expensive option. Always keep in mind that delta factors are not stable. They change whenever the price of the option premium and the underlying futures change, which is constantly. But these prices usually move in tandem, except when the option approaches expiration, causing its time value to decay rapidly.

There is at least one additional use for the delta factor. Occasionally you may want to hedge an options position against its underlying futures contract or hedge call options against put options. I'll go into more detail on strategies behind these trades later, but now I'll cover how the delta defines the hedge ratio.

First, I need to define "hedging." Hedging is the practice of offsetting the price risk inherent in any cash market (stocks, physical commodities, financials) by taking an equal, but opposite, position in the futures or options markets. The objective is to protect yourself from adverse price movement.

Here are two quick examples (I'll go into more detail when I discuss trading strategies).

In this first example, you own 1,000 shares of IBM stock. Your analysis indicates that it is due for a 20 percent price retracement. You don't want to sell the stock and buy it back at the "low," but you still want some price protection. You purchase an at- or in-the-money put. You ride the put down when the stock price falls, and you offset it at or near the low. Then you use the profits from this transaction to help compensate for the paper loss sustained by continuing to hold the stock.

For the second example, suppose you use copper or silver in your company's manufacturing process. Therefore, you follow the market very closely. You see a substantial rally ahead over the next 6 months. You buy enough calls to satisfy your production needs, with the idea of exercising these calls for long positions in the futures markets as you need the copper or silver. You can even take physical delivery on the futures positions, if you wish. Or you can offset the options as prices rise. Then you buy your needs in the spot (cash) markets, using the profits from the option to compensate for any increases in the cash prices.

The hedger is neutral—long in futures and short in cash, or short in futures and long in cash. By being on both sides of the market, the hedger freezes the price. The cost involved amounts to the transaction costs (commissions and fees) that must be paid to do the hedging. Some business executives think of hedging as price insurance.

When you use futures to hedge, you match quantity to contract because futures have a delta of 100 percent. If your objective is to hedge 40,000 pounds of live hogs, you use one futures contract. When you hedge with options, you have a different situation. Calculating the delta solves it. Your objective is to stay balanced or hedge-neutral. This ensures you have dollar-for-dollar protection. If the futures contract gains or loses, the delta of the option changes. Therefore, the delta tells you the proper ratio to maintain between the two to be neutral. If an at-the-money call has a delta of 50 percent, the hedge ratio is 2:1. To hedge the 40,000 pounds of live hogs, you would buy two option contracts, as opposed to one futures. This procedure also works for call versus put hedges. Here you want to match the call deltas (positive numbers) with the put deltas (negative numbers) and arrive at zero. For example, suppose you buy three calls with deltas of 40 each, or 120 ($40 \times 3 = 120$). To hedge with puts that have a delta of 60, you'd need to buy two calls ($60 \times 2 = 120$). The strike (exercise) prices and the expirations can vary, but the deltas must add up to zero to be delta- or hedge-neutral.

The delta factor can also be used to create a theoretical futures position. Futures positions have a delta of 100. If you wanted to get the same price action out of an option with a delta of 20 as you do from a futures, how many do you buy? The answer, of course, is 5 (100 divided by 20).

One key concept you must always be alert to is that deltas cannot be calculated just once or twice during the life of a contract. Once or twice a week, or even daily, may not be sufficient in very volatile markets. Whenever the price of the underlying entity changes, so will the delta. Now, small fluctuations won't seriously affect your hedge ratio, but large ones will. Some traders even plot 3-, 4-, or 5-day moving averages of delta, using each day's settlement price.

Get in the habit of figuring deltas regularly. Plot them on graph paper. Study them to uncover patterns. It's an excellent way of getting a feel for the relationship between an option and its underlying entity. In order to build on your understanding of the delta, there are a few other terms related to the delta factors you should be aware of as an option trader. The rate at which the delta changes (gains or loses) is known as the gamma. It is calculated in delta points. For example, if an option has a gamma of 10, for each point increase in the price of the underlying entity, the option's delta gains 10 points. If an option initially has a delta of 25 and the underlying option gains a point, the delta moves to 35. The value of knowing the gamma is that it will alert you to which options are most volatile. High gammas mean high opportunity—but also high risk. When trading for the first time, I suggest you avoid high gamma options because of the risk involved.

As mentioned earlier, options are wasting assets (see Figure 1–2). Eventually all expire worthless. As time runs out, the less attractive they are because their usefulness is evaporating. Think of them as a hundredweight of ice. The longer it sits, the more it melts. The more it melts, the smaller it gets. At some point, it is reduced to a puddle of water at room temperature. Then the water evaporates. Nothing is left. That's the life cycle of an option. The beta is the rate

F I G U R E   1–2

Options as Wasting Assets

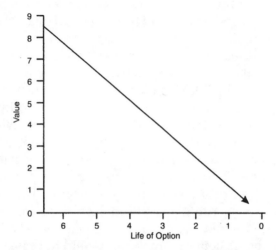

Options are wasting assets, because they eventually expire. Their time value
continually deteriorates as they move toward their expiration date. To make money
or conserve losses, you must offset or exercise options before they expire.

at which an option loses its time value. This rate increases as the
expiration date approaches.

Options prices are very sensitive to the volatility of the price
movement of the underlying entity (stock or futures contract). This
is such an important consideration that one of the upcoming chapters is devoted to it. But for now, I simply want to discuss the term
used to describe the relationship between changes in volatility of
the underlying entity and the price of the option. For whatever reason, the investment industry has not settled on a single name. It has
been called "kappa," "omega," "sigma prime," "vega," and "zeta."
In this text, I'll refer to it as "vega," which seems to be the term
accepted by most computer programs that calculate the theoretical
price of options.

I touched on carrying costs for futures and stocks earlier. One
of the most important of these is interest rates, more so for stocks
than futures contracts. An increase in interest rates decreases the
price (value) of an option because of increased carrying costs. The

term "rho" is used to describe the sensitivity of an options price to interest rates. It is always a negative number. When you begin to run theoretical options prices on some computer programs, you'll be asked for the interest rate. It will be used to calculate the rho as part of the equation.

One simple concept to understand, but a very key one, is market liquidity. Liquid markets are actively traded markets, where a lot of buying and selling is taking place. They are characterized by small price changes or narrow ranges between bid and ask prices. The opposite of liquid markets is not-liquid ones. The danger with not-liquid markets is that your order may not be filled or you'll get a bad fill.

Not getting filled on an order can mean missing an opportunity or not being able to exit a market at a profitable position. A bad fill is simply a price worse than you expected. For example, you want to buy a call. You place an order to buy a call at the market when it is trading between 10 and 15. You get filled at 15 or 16, the high for the day. A better fill for a call would have been around the low of the day, or 10. If each point is worth $100 in this particular market, the difference between 10 and 16 would be $600.

You find out about liquidity by checking the daily volume for each market you trade. It is published in financial newspapers, announced on cable television programs, broadcast over the Internet or over an electronic price quotation service. Your broker should have access to all of these. Daily volume can range from less than one hundred to well over a hundred thousand.

You should always discuss liquidity with your broker before placing an order. Be sure to check the volume of the specific strike price you are considering. Strike prices at- or in-the-money normally have high daily volumes, much higher than out-of-the-money strike prices. Deep-out-of-the-money options can have little or no volume on any given day.

Exchange rules come into play with liquidity. Some exchanges have a "not-held" rule. This means the floor traders are not held responsible for filling every order. In active markets for stocks and futures, any order that's "at-the-market" gets filled. If nothing else, a broker or brokerage firm that is designated as a market maker will take an order that is having problems finding a home. But with some of the not-liquid options markets, the exchanges protect their floor brokers from having to take a sure loss. This means you could place an option order only to have it come back "unable," even if the bid-ask price is at or around your price.

This normally would not happen if you were trading stocks or futures. You could get an unable ("unable to fill or execute") in these markets, but only if your price was out of the trading range. If you placed an order, for example, to go long cotton at 60 cents a pound or better when it was trading at 50 cents, no one would fill your order because cotton has a 2-cent daily trading range. It cannot run up 10 cents to pick off your order. The market is out of range at this point, and the order would be returned as "unable." Option orders can also be returned from the trading pits as unable if a buyer or seller cannot be found.

In the futures markets, there can be limit-up and limit-down days. These are trading sessions when a particular futures contract is trading beyond the daily limit permitted by its exchange. For example, the daily limit for the Chicago Mercantile Exchange's live hog contracts is $0.015 or $600 (40,000 pounds × $0.015), or 150 points. This means that if the contract is bid up $600 in a single trading session, trading will be halted—this is similar to what happens on the stock exchanges when volatility gets out of hand. For the record, the hog contract has traded limit up for as many as 6 days consecutively. This is one of the serious risks faced by futures traders.

When limit-up days occur, option prices move wildly because of the high volatility. A put option which is in-the-money by a penny ($400) or two ($800) quickly loses all intrinsic value. The time value decreases as well because the trend is so violently in the opposite direction. As you guessed, all the in-the-money calls gain value as fast as or faster than the puts lose. Competition for calls becomes fierce, moving them into the oversold category.

Since I mentioned market orders and "or better" orders, there are a few other general ones you should be aware of as long as I am discussing terminology. Specific types of orders are used by your broker to accomplish your trading objectives. You don't necessarily need to know each type, but being aware of them helps you plan your strategies. Most traders explain what they want to accomplish with a trade and have their broker decide on the proper order. But you need to know all your alternatives to intelligently plan your trades.

There are basically two types of orders. Those that have no restrictions on them and those that do. Market orders have no restrictions.

When floor brokers receive market orders, they immediately execute them. In fast-moving markets, you could call a market order to your broker, who would relay it to the order desk on the exchange trading floor. Your broker could get your fill back so fast that you could conceivably still be on the line with him. Speed is the advantage of using market orders. Floor brokers give them the highest priority. You're virtually assured, barring limit-up or -down days, of being filled.

The disadvantage of market orders is that there is usually no time to stop or change them. Once in the pits, they'll get filled. For example, you're trying to buy a call priced at 4 cents or $4 per unit (bushel, barrel, hundredweight, share, etc.). Your fill comes back at 6 cents. The 2-cent difference occurs due to the speed at which the market is moving. A 2-cent difference or a nickel or whatever can sometimes mean the difference between a winning trade and a loser.

The second type of order has some time or price restriction attached to it. Limitations can be attached to the time the order can be executed or the price. For example, there are orders that can only be executed at the opening (market-on-open, or MOO) or closing (market-on-close, or MOC) of a trading session. Each exchange established a specific opening (first 15 minutes) and closing (last 5 minutes) period. Certain orders must be filled right away (fill or kill) or canceled. Others are good for various periods of time (good till canceled, day orders, good through/time date).

There are a variety of orders that include some type of price limitation. I've already alluded to the "or better"-type order, where the floor broker is obliged to meet or beat the price you specify. Then there are stop orders. A stop is a limit, but once reached, the order becomes a market order. A market-if-touched (MIT) works about the same way as a stop order. Your order is executed if a certain price is touched, but the price you receive doesn't have to be better than the touched price.

The advantage of price limitation is that it prevents your order from being activated until certain conditions are met. Traders often use price limitations in conjunction with their trading system or the forecast of what the market they are trading is expected to do. Or they use them to get a certain price. Much of this type of thinking will become clearer when I discuss price trend forecasting, trading systems, and strategies in the chapters ahead.

The big disadvantage of placing limitations on your orders is that they may not be filled. An "or-better" order sounds good, except when you miss a profitable opportunity because the floor broker could not execute it.

The last type of order is a discretionary order. You can relinquish the trading discretion in your account to your broker. This can be unlimited, in the case where you sign a limited power-of-attorney for your broker to trade on your behalf; or it can be limited. Limited discretion gives the broker the right to enter a trade in the market based on your prior approval of certain time or price considerations.

For example, you discuss your current position with your broker in the evening. You will be traveling the next day and out of touch. You decide that you want your position closed out if the market opens higher (or lower or reaches a certain price). The broker only has discretion to make this trade on your behalf. When you give price discretion, it is usually best to place some upper and lower limits on the discretion.

This is by no means an exhaustive discussion of orders. The purpose is only to make you aware of the alternatives you have to placing a simple market order when buying or selling puts or calls.

When you first consider trading options, be prepared for what I call "options shock." It is sort of like futures shock in that you initially may feel overwhelmed. If, for example, your investment background is commodities, you're generally familiar with 40 or so actively traded commodities. There are four to twelve contract months for each, and you can go long, go short, or spread them.

Let's take something simple, like corn, as an example. It has five delivery month contracts (March, May, July, September, and December). Trading it long or short gives you two other choices for a total of ten. With options, you additionally have to select from at least six strike prices, each with both puts and calls available. Then you can be a buyer or a seller.

Stock traders face a similar maze. There are thousands of stocks with options to choose from, multiple strike prices, and puts and calls for each. Experienced stock traders additionally need to become comfortable trading in downtrending markets utilizing puts or selling calls.

If this all weren't confusing enough, there are options to consider besides the classic option-on-futures or option on 100 shares of over 7,000 stocks. There are index options of all sorts:

S&P 500 or 100

Utility Stock Index

Japanese Stock Index

OEX

Value Line Composite Index

Gold & Silver Index

Computer Technology Index

Institutional Index

Major Market Index

Oil Index

National Over-the-Counter Index

Commodity Research Bureau Index

Goldman Sachs Commodity Index

Growth or Income Mutual Fund Indexes

And there are foreign currency options from around the world—Australian dollar, British pound, Canadian dollar, deutsche mark, French franc, Japanese yen, Swiss franc, European currency unit. The currencies can be either European or American style, depending on which exchange you trade. Plus they can be full or half-size contracts, again depending on the exchange. If this is not enough of a selection, the currencies also offer cross-rate options, like a deutsche mark–yen cross.

How long an option do you want to trade? You can trade Long-Term Equity AnticiPation Securities (LEAPS) on individual stocks. These give you buying or selling rights on the underlying security for up to 2 years. Or you can trade 5-day options in the precious metals.

Additionally, you must keep tabs on the day your options expire. There are no uniform methods among the futures exchanges for assigning expiration dates. Some expire on Thursday, others on Friday or Saturday. Most expire 4 to 6 weeks prior to the delivery date of their underlying futures contract. Most stock options expire on the Saturday following the third Friday of the expiration month and are set up on a quarterly cycle. Always check with your broker.

Also, make sure your broker knows the rules of the exchange you'll be trading on regarding the expiration of an in-the-money option. On some exchanges if an option expires in-the-money, any floor broker can exercise either side. You could unexpectedly be given a losing trade. Keep your broker on his or her toes.

My point is that there is an extraordinary number of options and option combinations to evaluate. For new traders, I recommend trading options-on-futures over an intermediate period of time, for example, from 30 to 60 days to expiraton.

The rationale for this suggestion is that these option markets are liquid and are reasonably easy to understand. Further, I suggest new traders to seriously consider some of the options traded on the MidAmerican Exchange because they are options on half-size contracts. It often pays to learn to walk before you try to run.

CHAPTER 2

Learning the Basic Option Trading Strategies (and Uses) to Solidify Objectives

"The happiest time in any man's life is when he is in red-hot pursuit of a dollar with a reasonable prospect of overtaking it."

Jack Billings (1818–1885)
American Humorist

Key Concepts
- Power of Leveraging
- Law of Probability and Three Rights
- Distribution of Winning and Losing Trades
- Four Ways to Approach the Market
- 12 Trading Strategies
- Writing Puts and Calls
- Hedging with Options

Normally a discussion of strategies comes at the end of a book of this nature. I opted to place it up front because I want you to be thinking about your strategies and objectives as you study the remainder of the text. My expectation is that if you have a good idea of what you want to gain from option trading, all the information that follows will be more meaningful.

Options come in all sizes, shapes, and forms. Some are built for speed and daring; others for dependability and smooth sailing. Still others are very crafty, putting the odds in your favor. Which suits you?

There are basically two types of option traders, namely speculators and hedgers. The speculators ("specs") are risk takers, while the hedgers wish to transfer risk to someone else. The objective of the specs is to generate profits. Hedgers want to preserve their gains or protect positions. I'll probe the role of speculators first, since this is what most traders are. Then I'll discuss hedging, because it can have a dramatic impact on the specs. Some investors, who speculate in one market, such as stocks, manage or hedge their risk in the options markets.

All speculators have the same goal (profits) even though their strategies vary widely. Also, the complexion of the market can strongly influence your approach. You must be more flexible than the markets you trade to win.

As you develop your strategies, keep two things in mind—the *power of leveraging* and the *law of probability*. Leveraging allows you to control a large amount of an asset with a small amount of capital. For example, let's say corn is trading at $2.10 per bushel and the $2.10 strike price call is trading for $0.055 (5½ cents). The contract has 6 weeks to expiration. You would have control of $10,500 worth of corn with an option that costs $275 plus $50 in commissions and fees for a total of $325. Your leveraging ratio would be 32:1 ($10,500/$325). A similar situation in the stock market would be the purchase of an AT&T $40-per-share call at 1⅞ ($1.875), or $187.50 (100 shares × $1.875), plus $40 of transaction costs. If the stock is at 41¼, or $41.25 per share, at the time, the total value of 100 shares would be $4,125. Dividing this by the total invested gives you a ratio of approximately 18:1 ($4,125.00/$227.50).

Now, what does leveraging do for you? It gives you the opportunity to make a high return percentagewise on your investment. Take the corn example. When corn gains a nickel, the total value of the contract increases by $250. Your $2.10 option is at-the-money. Therefore, its intrinsic value would increase by an equal amount, or a 76 percent return to the premium. Additionally, there could be some increase in the time value, depending on how much is left to expiration.

You would have a similar situation with the stock option. It's for 100 shares. Therefore every $1 gain per share increases the value of the in-the-money option by at least $100 of intrinsic value. A gain of $2 per share almost doubles your investment.

Also, remember that these are relatively short periods of time. These options had only 6 weeks to expiration. There are few investment opportunities that offer the potential of doubling your money in a matter of weeks or months. And it is common for the moves to occur just before expiration. To make the gains meaningful from a dollar standpoint, serious option investors trade multiple lots of contracts. This would be 10, 20, 50, or 100 options contracts (lots) at a time.

Before you call your broker to make one of these "piece of cake" trades, you need to think it through. The sellers of these options, in the trading pits and all over the world, are not dumb. When you buy a put or a call, someone has to sell it to you (underwrites it). That person expects your option to expire worthless, so he or she can keep the premium. To put it more bluntly, the option seller wants you to lose 100 percent of your investment.

Writing (selling or granting) options entails more risk than buying options. The seller can be assigned the opposite side of the buyer's option position any time the buyer decides to exercise it. It would probably only be done when it was in-the-money, which means the seller would be losing money as soon as the buyer acted. The loss could be substantial. Think what would happen to a seller in the futures market who has been assigned a position that is making daily limit moves against her. For hogs on the CBOT, it would be $600 per day per contract. After a day or two of limit moves, the daily limit— or the maximum a contract can trade in a single trading session— might be extended to $700 or $800. Some markets, like the foreign currencies, have no daily limits. It is because of this possibility that the downside risk of selling options is considered unlimited. The upside or profit is limited to the amount of the premium, less transaction costs.

Why would anyone enter an investment where the profit potential is limited and the risk of loss is unlimited? This will be covered when I discuss selling options as a strategy. You'll also learn why you may have a better chance of making money in the long run selling than buying options. Your first clue is this: Most options expire worthless!

Since selling options can be a profit strategy, the insinuation is that (Figure 2–1) buying options must be risky. It isn't a piece of cake.

F I G U R E 2–1

Breakeven Analysis (call)

Buying an At-the-Money Call in Futures Market

Assumptions:

Traders outlook is bullish for gold.
COMEX $400.00 call priced at $12.00 per ounce.
Commissions and fees amount to $75.00.

Premium ($12.00 × 100 oz.)	$1,200.00
Commissions and Fees	75.00
Total Investment	$1,275.00

Breakeven—Gold must increase $12.75 per ounce.

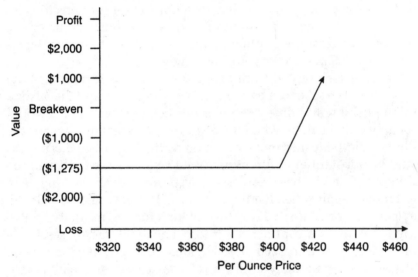

Note: Analysis does not take time value into consideration, since it is virtually impossible to calculate with any exactitude.

The risk may be limited to your initial investment, but all of it can be lost. This line of thought brings us to the law of probability. The strategy you select will have a greater probability of succeeding if it is diversified. This means trading options on a variety of commodity markets or stocks. To be a successful trader, you must be in the right market, at the right time, and on the right side (put or call). You are investing in what you expect to happen in the future—5 days, 5 weeks, or 5 months from the moment you call your broker and place your order. You have no way of knowing if you will be right or wrong until afterward.

At the same time, you must protect your risk capital as carefully as possible. This requires sound money management techniques, which is the second part of the law of probability. A sure way of ending a trading career abruptly is putting all your risk capital on one or two trades. A "double-or-nothing" approach to the markets will invariably—in my opinion, which is based on a lot of experience—lead to you ending up with nothing.

Just as you increase the probability of being in the right market at the right time on the right side by trading several different markets, you'll increase your probability of success by putting no more than 10 percent of your risk capital on any one trade. This gives you at least 10 opportunities to select one or more trades to pay for the ones that lose or just break even.

You must further consider what is known as the "distribution of winning-losing trades." For example, your distribution might look like this:

Trade distribution:
> Ten trades executed
> Five winners
> Five losers

Of the five winners:
> Three small or breakeven
> One modest size
> One big or decent size

Of the five losers:
> Four small
> One modest

Your objective is to let winners run and cut losers short. If you do an analysis of the trading performance of the most successful CTAs (professional commodity trading advisors), you rarely find a winning percentage higher than 60 percent.

Keep in mind, the prime objective is to make money, not to generate a high percentage of profitable trades. You can be a net winner with a low percentage of winning trades. By low, I mean 40 percent, 30 percent, or even 20 percent.

To do this, you must be very disciplined. You develop, for example, strict rules for exiting losing trades (these are detailed later). For instance, when a trade begins to make money, you place a trailing stop behind it (below a long position or above a short). Eventually this stop position is taken out, offsetting your position. In other words, the market decides for you when to close a position.

A word of caution is necessary regarding trailing stops in some thinly traded markets. Thin markets are markets with low volume. If you or your broker checks the volume of trades on a daily basis and determines the market you're in is thin, don't actually put your stop in the market. Have your broker closely follow the market and close out your position if it trades where the stop would have been placed. The reason for this is that in thin markets ("scalpers") floor traders sometimes trade down to where stops are, pick them off, and then go back to where they were trading. What is happening is floor traders see the stops, short the market until the stops are hit, and then go long again. You're out of the market at a loss for no reason except a smart floor trader took you out.

Trading several markets takes more of your time, as does updating a system that is tracking a number of alternative commodities or stocks. On the one hand, you need to be diversified. Equally important is to avoid becoming overextended financially, psychologically, or personally.

OVERVIEW OF STRATEGIES

I begin with a general discussion of option trading strategies and then move to an explanation of specific ones. There will be no attempt to provide a definitive discussion of every possible strategy. But most of the common ones will be covered.

There are four basic ways to approach the option markets. First, you can analyze the market you wish to trade and forecast the trend. Or second, you can use a system that mechanically alerts you to options that are overpriced or underpriced. Third, you can be a writer or grantor of options and underwrite the buyers. Last of all, you can use a combination of the three approaches just mentioned.

As a market analyst, you can use fundamental or technical analysis to decide where the market is headed. Fundamental analysis is the study of all the underlying factors that affect the supply-demand equation for a stock or commodity, and thus the price.

Technical analysis disregards all fundamental factors. It relies on the study of price action, rates of change in price, volume, or open interest. Prices are often charted and the patterns analyzed. Or the analysis could be done on a computer. Numerous studies would be run with the hope of discovering recurring patterns. These two alternatives will be discussed in detail in later chapters.

Once you arrive at an opinion of the direction of the trend of a market, you must quantify it. Are you mildly or aggressively bullish or bearish? Are you neutral—do you think the market will trend sideways?

Your second possibility is to adopt some type of mechanical trading system. The most common of these are computer programs (see Appendix 4 for a listing and descriptions) that evaluate options using a theoretical pricing model. You calculate the price of an option using the amount of time to expiration, market volatility, carrying costs, and current price of the underlying entity (stock or futures contract). Then you compare this theoretical price with the actual market price. This tells you if the option is overpriced or underpriced. You then buy or sell a call or a put depending on the analysis. Since volatility is the key factor in the analysis, I'll talk in more detail about how this is calculated in chapter 5 on volatility.

The third approach is to be a writer of options. Here you have limited profit potential and substantial risk. Again, you have some choices, such as whether you sell covered or uncovered options. Underwriting options is an approach that is rarely recommended to new traders, but it can have an appeal to experienced investors—even if they are new to options.

Lastly, it is common to create trading strategies that utilize a combination of the four basic methods. You can, for example, use the premium you receive for writing an option to buy one. This is sometimes called a "free" trade since you don't have to pay directly for the option you buy.

SPECIFIC TRADING STRATEGIES

I'll begin with "Raging Bull" strategies. You think the market is grossly oversold—poised for a major move higher.

Let's use a futures contract in the silver market as an example. It has been trading below $5.00 an ounce for almost 2 years and under $4.00 for the last 12 months. Your analysis indicates that it is about to make a move from the $3.50 level to over $5.50. Your best estimate is silver at $6.00 or better within the next 60 days.

The simplest and most common approach is to buy a call. The call gives you the right, but not the obligation, to take a long position in the underlying futures. As the underlying futures gains, so does the call. Either you can exercise the option and take your long futures position, or you can offset your option positions at a profit as it gains intrinsic value.

The more difficult question is which strike price to buy. Here are your choices for calls with 60 days to expiration on 5,000 troy ounces of silver:

Call-Strike Prices	Price/Ounce
$3.00 55 cents	In-the-money by 50 cents
$3.25 30 cents	In-the-money by 25 cents
$3.50 5 cents	At-the-money (current silver price)
$3.75 3 cents	Out-of-the-money by 25 cents
$4.00 1 cent	Out-of-the-money by 50 cents

The $3.00 strike price option has 50 cents intrinsic value and 5 cents time value. The $3.25 strike price is a quarter in-the-market with a nickel time value. The at-the-market option has no intrinsic value and 5 cents time value. The two out-of-the-money options only have time value. The farther out-of-the-money, the less time value.

Studying these evaluations indicates the market doesn't agree with your analysis. Sellers of out-of-the-money options are making them very attractive. A $4 call carries a $50 premium (5,000 oz. × $0.01).

The at-the-money strike price is very reasonable if your analysis proves to be anywhere near accurate. At 5 cents per ounce, the call costs $250 plus transaction costs of, say, $50, for a total of $300. A $2-per-ounce gain in silver would amount to $10,000. The ratio of profit to loss would be 33:1.

Another way of looking at this trade would be to calculate the breakeven point. How much does silver have to gain just to get your $300 back? Each penny gained adds $50 to the intrinsic value of the at-the-money option. Therefore, all that is needed is 6 cents and maybe a little less, if the time value increases some as well.

A put works the same way, but in the opposite direction. I call it the "Doomsdayer's Strategy." You think, for example, that the shares of ABC Company, now at $50, are headed south in a big way, maybe even as much as 50 percent. Checking the paper, or calling your broker, provides the following prices:

Put-Strike	Prices	Price/Share
$40	¾	Out-of-the-money by $10
$45	1	Out-of-the-money by $5
$50	1⅛	At-the-money (current share price)
$55	7	In-the-money by $5
$60	10	In-the-money by $10

Again, the market isn't bracing for a major slide in the price of ABC's stock. It is only asking $75 (100 shares × $0.75) for an option with a strike price of $10 out-of-the-money and $100 for one $5 out.

The breakeven for the at-the-money option would be a decrease in the per-share price by approximately $1.00 plus ⅛ in time value, or about $1.50 to include transaction costs. The reward-to-risk ratio, if the shares do drop $25.00 per share before the at-the-money option expires, would be approximately 16:1 ($2,500/150). (See Figure 2–2.)

I call these two strategies the "Raging Bull" and "Doomsdayer" because that is the way they are often presented to investors. For these strategies to work, you need a major price move, which is rare. Additionally, you really have only one opportunity to profit and that is if your analysis is absolutely correct. Experience has shown that the odds of this happening consistently are remote.

One of the most expensive errors people new to options make is getting caught up with the excitement a broker may create regarding a trade. "Silver is headed off the charts! Get in now! What can you lose?" The answer is 100 percent of your investment.

This trading strategy requires you to hit home runs each time at bat. If they pay off big, they are grand slammers! It is for this reason, in my opinion, most option traders, particularly new and/or small traders, are net losers as option traders. There are occasions when you should swing for the fence, but most of the time you should try a strategy with a higher percentage success rate. Unfortunately, most

F I G U R E   2–2

Breakeven Analysis (put)

Buying an At-the-Money Put in Stock Market

Assumption:

Trader's outlook is bearish for stock.
ABC's stock's $50.00 per share puts are priced at $4.00 per share.
Commissions and fees amount to $50.00.

Premium ($4.00 × 100 shares) $400.00
Commissions and Fees   50.00
Total Investment $450.00

Breakeven—ABC stock must decrease $4.50 per share.

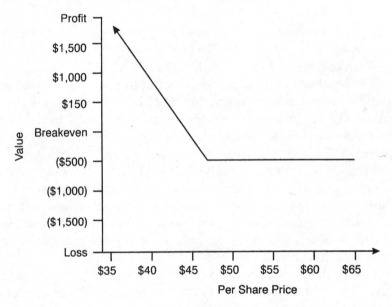

stock and futures options traders never get beyond this point in the
learning curve.

BASE HITS

Attempting to get walks, singles, or doubles consistently is a much
higher percentage play—consider the spread trade. Spreading simply means trading both the long and short sides of a market at the

same time. The spreader attempts to make a profit by anticipating the amount of change in the price movement between options. The price differential between the two markets is the spread, thus the name.

There are various types of spreads. They can be described by market (exchange), commodity, or delivery month. You can trade "inter" or "intra" types of spreads or combinations of the two. "Inter" means between two different entities; "intra" means within the same entity. See Figure 2–3.

VOLATILITY SPREADS

Volatility spreads are one of the simplest and most useful options spreads to learn and use. The spread trader buys a put and a call at the same strike price. For example, suppose corn is trading at $2.52 per bushel. You buy a put and a call at a strike price near- or in-the-money, at, say, $2.50.

You do this with the expectation that corn prices will become more volatile. For example, this situation can occur at the time of the release of an important crop report, such as "Farmer Planting Intentions." You may not be sure if the report will be bearish or bullish, but you expect it to cause a lot of excitement—therefore, price volatility!

This situation can occur in the stock market when there is news of a takeover, a new product introduction, or a major stock split. If this happens, the premium value of both your options could increase. Or if the market takes off north (bullish), your call becomes more valuable. If it goes south, your put goes into the money. Your net position should improve one way or the other.

F I G U R E   2–3

Types of Spreads

Long Corn/Short Wheat	Intercommodity
Long CBOT Wheat/Short MN Wheat	Intermarket/Exchange
Long July/Short Sept.	Interdelivery
Short CBOT Corn/Long CBOT Corn	Intermarket/Exchange
Short July Corn/Long Sept. Corn	Intracommodity

If the impact of the news is already in the market with prices and volatility unchanged, you lose. You must then decide whether to hold your spread in hopes of future volatility or take your loss. A little added advantage of this spread is that the premium is usually lower than the sum of the premium costs of both a put and a call purchased separately. A put-call spread purchased at the same time at the same strike price is called a "special double," and it will be discussed as a separate strategy later.

TIME OPTIONS

Another popular option spread is the time spread. It involves options with the same strike price but different expiration dates. The prime objective of a time spread is to take advantage of the tendency of the time value of an option to decline at a very rapid rate before finally disappearing just prior to expiration. Typically, a put or call with a nearer expiration date is sold (written), while a put or call with an expiration date that is more distant is purchased. The option sold and the option bought both have the same strike price.

The intent is to sell time. The more distant put or call loses its time value at a slower rate, which limits the risk of such a position by removing the potential for unlimited loss on sharp upside moves.

THE PRICE INFLUENCING STRATEGY

The price of the underlying entity is the prime determinant of how much an option spread will profit. If the spread position is established by selling the nearby option and buying the more distant one when both options are at-the-money, the spread will be profitable if the futures or stock prices remain relatively constant. It will tend to lose money if the prices change.

The spread will be at its widest when the price of the underlying contract is closest to the exercise price of the option. The spread will narrow as the price moves away from the exercise price in either direction.

BULL SPREADS

Spreads can also be described as bull or bear spreads. A bull spread is usually long the nearby delivery month and short a more distant delivery month. Or you could buy an option at a price close to the current price of the underlying entity and sell one that is at a higher price. The expectation of the bull spreader is that if the price of the underlying entity rises, the effect will be felt strongest by the closest delivery month or the option closest in price to the underlying option.

Let's say May silver is now at $4.50 an ounce and a $5.00 strike price (50 cents above an out-of-the-money) call has a 25-cent premium, or $1,250. But you really don't want to take that much of a risk by simply buying a call. Silver has been somewhat lackluster over the last few months. You feel that it is headed higher; however, you're not sure of the timing or how far it will go. On the other hand, you don't totally want to miss what you think is an opportunity.

An alternative would be to write (sell) a call with a higher exercise price and buy the nearby option. In this case, the $6 silver call could be sold for about 14 cents, or $700. The spread between the two options would be an 11-cent debit. Such a spread is called a bull spread, and the risk in the position is limited to the amount of the original debit.

Risk limitation does not occur without giving up opportunity. If the risk is limited, so is the profit potential. The maximum profit in a price spread is calculated by subtracting the original debit from the difference between the exercise prices of the two options. Here's the mathematics:

Difference between options	$1.00
Debit	0.11
Maximum profit/oz.	$0.89
Maximum profit (5,000 oz. × $0.89) =	$4,450.00

In this example, the debit or net premium cost is 11 cents and the difference between the exercise prices is $1. You are risking 11 cents to make $1. At the expiration, a silver price of $6, for example, would mean the $5 call would have gained $1 per ounce, or $5,000, while the $6 call you wrote would be exercised. This means you would lose the premium of 14 cents, but you would also be closed out in the market. You would be long and short one silver contract. If the price of silver dropped below $5, both options would expire worthless, and you would experience the maximum loss of 11 cents.

BULL SPREAD RULES

The rules concerning option bull spreads that you must keep in mind
are as follows:

+ A bull spread profits only when the spread widens.
+ The maximum profit on the bull spread is the difference
  between the strike price minus the debit (the amount of the
  cost of the long side exceeds the proceeds or the premium
  received for selling the short side).
+ The maximum loss is the debit or the net premium paid.

Now I'll discuss the bear side of spreads.

BEAR SPREADS

Basically, bear spreads call for an opposite market opinion to bull
spreads. You use them when you believe the market is about to move
lower.

If you project a declining silver market, you could sell (write)
a put at a lower strike price option, e.g., $5.00, and buy one at a high-
er one, say, $5.50. At the time the options expire, if the price of silver
never goes below $5.00, the put you write will not be exercised by
the person who bought it—you keep the premium. The put you
bought at $5.50 may gain, depending how low silver goes. At $5.00
silver, at expiration, your $5.50 put has 50 cents intrinsic value, or
$2,500. Additionally, you would pocket the premium of the $5.00
put you underwrote. This would be your maximum profit.

If silver continued to trade lower before expiration of these options,
your profit would not increase. The reason is that the owner of the
option you sold would exercise it. To protect yourself and end your
market exposure, you exercise the put you bought. You would be long
and short the silver market or neutral (one position offsets the other).

The rules for bear option spreads are as follows:

+ A bear spread profits when the spread narrows.
+ The maximum profit on a bear spread is the credit (the
  amount the proceeds from the short side exceed the cost of
  the long position and the transaction costs).
+ The maximum loss on a bear spread position is the difference
  between the strike price less the credit and transaction costs.

BULL AND BEAR TIME SPREADS

Just as you can trade bullish or bearish price spreads, you can trade bullish or bearish time spreads. A bull time spread would involve buying a deferred month and selling the nearby month.

If, on the other hand, you were bearish, a bear time spread would entail buying the nearby expiration month and selling a deferred expiration month—for example, buy a July ABC Company call option and sell an October call option. The rules cited earlier apply to bull and bear time spreads, as well as bull and bear price spreads.

STRADDLE STRATEGIES

The straddle trade is similar to a spread in that you trade both sides of the market at the same time, but there is a big difference. A straddle is a type of spread that entails the purchase of a put and a call (called a "long straddle") or the sale of a put and a call (called a "short straddle"). Previously, the examples involved buying and selling either puts or calls, not combinations.

Earlier I concentrated on bullish and bearish trading strategies. These are valuable tools when you know (or think you know) where the market is headed. What do you do when you think the market is going to take off, but you don't know in which direction? Or even more puzzling, how can you trade a trendless market?

The answer is the straddle. The long straddle, like the volatility spread discussed earlier, is a trading strategy to take advantage of dramatic market moves, even when you're not sure of the direction.

THE LONG STRADDLE

In its simplest form, the long straddle involves the purchase of a put and a call for the same stock or commodity that share the same expiration date and strike price. To be successful, a big market move must occur before the options expire.

Let's use the corn market again as an example. The primary growing region is the Midwest. In this scenario, the Midwest has been in a drought for the past 2 years. As planting time approaches, the subsoil moisture has not been replenished. Last year the crop got by with a series of timely rains. Just enough rain occurred when it was desperately needed. The year before, the farmers were not as lucky. What will happen this year?

The United States rarely has two down years of corn production in a row. Additionally, the price of corn is very price sensitive, often referred to as "elastic." When supplies are scarce, the price skyrockets. When supplies are plentiful, the price plummets. When supplies are adequate, the price moves sideways.

This year you think one of two things is going to happen— either there will be plenty of rain or there will be a drought. A full measure of rain will wash out prices. A drought will fire prices to new highs. How do you trade these fundamentals?

If you think the price range could be anywhere from a low of $1.40 to a high of over $3.00 per bushel, you might decide to use a long straddle. You'll buy a near-in-the-money put and call at, say, $2.40. See Table 2–1.

As you can see, the money is made with this strategy on the extremes. The more the market declines, the more valuable the put becomes. The opposite is true of the call. At some point before expiration, the savvy trader would abandon one leg and exercise the other. The point is, you're in good shape to take advantage of a bull or bear run without knowing which will occur beforehand!

T A B L E   2–1

Long Straddle

Future Price	Call Price	Put Price	Total	
1.40	(20.0)	80.0	60.0	
1.60	(20.0)	60.0	40.0	
1.80	(20.0)	40.0	20.0	
2.00	(20.0)	20.0	0.00	
2.20	(20.0)	0.00	(20.0)	
2.40	(20.0)	(20.0)	(40.0)	Max. Loss
2.60	0.00	(20.0)	(20.0)	
2.80	20.0	(20.0)	0.00	
3.00	40.0	(20.0)	20.0	
3.20	60.0	(20.0)	40.0	
3.40	80.0	(20.0)	60.0	
(In Dollars)				

You lose, of course, if the market trades trendless from the time you enter until expiration of the options. The maximum you lose is the two premiums paid for the put(s) and call(s) and the associated transaction cost. The upside is theoretically unlimited. I say "theoretically" because all bull or bear markets eventually end.

THE SHORT STRADDLE

The opposite trading strategy to the long straddle is the short straddle. Rather than going long a put and a call, you short or sell a put and a call. This is the ideal strategy to use when you think the market is going to be flat and lifeless for the duration of the options sold.

Let's use the same corn scenario described above. In this case, you think the rains will be normal, crop production will be normal, and the market will trend sideways until our options expire. Rather than buy a $2.40 call and a put, you sell them. See Table 2–2.

The short straddle is attractive because it allows you to take advantage of the normal decline in an option's time value as it approaches expiration. In this case neither option is exercised, which means you retain the full premium on both sides. The danger is a serious price move, which could be very expensive.

T A B L E   2–2

Short Straddle

Future Price	Call Price	Put Price	Total	
1.40	20.0	(80.0)	(60.0)	
1.60	20.0	(60.0)	(40.0)	
1.80	20.0	(40.0)	(20.0)	
2.00	20.0	(20.0)	0.00	
2.20	20.0	0.00	20.0	
2.40	20.0	20.0	40.0	*Max. Gain*
2.60	0.00	20.0	20.0	
2.80	(20.0)	20.0	0.00	
3.00	(40.0)	20.0	(20.0)	
3.20	(60.0)	20.0	(40.0)	
3.40	(80.0)	20.0	(60.0)	
(In Dollars)				

As with all trading strategies, you begin with a strong opinion of where the market is headed. Always be prepared to adjust your position(s) whenever you discover that all or part of your underlying analysis is changing or is incorrect.

STRANGLING THE MARKET

The "strangle" is a type of straddle. It differs from the simple straddle just discussed in that both legs do not share a common strike price and are out-of-the-money. They are similar in the sense that you are trading both sides of the market and you are trading a long or a short strategy.

Another critical characteristic is different: the return curve. With the simple straddle, the maximum return was available at only one price point. Refer again to Tables 2–1 and 2–2. With the strangle, the maximum return is available at several price points. See Table 2–3.

Strangles also differ from straddles when it comes to the usual position taken by traders. The short strangles are more commonly used than straddles, whereas you'll usually find straddles to the long side.

TABLE   2–3

Strangle

Future Price	Call Price	Put Price	Combined Total
66	$1500	($4500)	($3000)
68	$1500	($2500)	($1000)
69	$1500	($1500)	0
70	$1500	($500)	$1000
72	$1500	$1500	$3000
73	$1500	$1500	$3000
74	$1500	$1500	$3000
76	($500)	$1500	$1000
77	($1500)	$1500	0
78	($2500)	$1500	($1000)
80	($4500)	$1500	($3000)

WHY USE A SHORT STRANGLE?

The basic strategy is the same for the short strangle as it is for the short straddle or the volatility spread. You see a market that you expect to trade flat until the expiration of the options. The strangle strategy becomes particularly attractive if there is a history of unexpected volatility in the market you plan to trade, or perhaps some seasonal patterns that might come into play. The short strangle, because of the reasons mentioned earlier, insulates you more from this kind of threat than does the short straddle or volatility spread.

The short strangle finds favor with traders over the short straddle because it further reduces risk. To begin with, you are trading different strike prices and they are out-of-the-money. Therefore, the underlying contract must move farther than with a short straddle, which has the same strike price and is in-the-money or near-in-the-money, before someone will exercise one of the options you write.

As with most investment strategies, less risk means less reward. This is true here since the premium for out-of-the-money options is comparatively low. Here's an example. Your analysis for this scenario indicates the bond market will be stable for the next 6 to 9 months. Therefore, you enter a short strangle by selling an out-of-the-money call at 74 and an out-of-the-money put at 72. The underlying futures contract is trading around 73. To simplify this example, suppose both options have the same premium, $1,500. The premium you receive is $3,000, less any transaction costs. Table 2–3 demonstrates a variety of market price changes and their impact on your investment.

Notice how the maximum income increases from one price point on the straddle tables (Tables 2–1 and 2–2) to three price points on the strangle table (Table 2–3), specifically between the 72 put and the 74 call strike prices. Between these two points, both options are still out-of-the-money. When both options expire out-of-the-money, you retain the full premium. This, of course, is the maximum profit you could expect. If one of the legs of the strangle expires in-the-money, then your profits diminish proportionally.

Prices can, at least theoretically, increase indefinitely and can decrease to zero. When prices move dramatically against one leg of the strangle, the option represented by that leg is likely to be exercised. This means you'd be assigned an equal and opposite position in the futures market. For example, if you are short a put, you will receive a long futures position if the option is exercised. Since the market is moving against you, it could quickly become expensive.

CALCULATING BREAKEVEN

For this reason, strangle strategists often calculate their breakeven point. It is slightly different depending on the direction of the market. For example, if prices go over the call strike price, it is likely the short will be exercised.

But it is only when the loss exceeds the net premium received that you become a net loser. You can define the breakeven point by adding the call strike price to the net credit (premium received). Our 74 call strike plus the $3,000 premium, or 3:00, adds up to 77:00 as breakeven.

The downside breakeven is determined by subtracting the premium from the put's strike price. Once this price is exceeded, you begin to lose money. In our example, you would reduce the 72 put by 3:00, which comes to 69:00.

THE LONG STRANGLE

Most of the preceding section was devoted to the short strangle as opposed to the long strangle. My reason is that a long strangle is rarely used. Everything mentioned about the short strangle is true for the long—only in reverse.

THE LONG SYNTHETIC

In options trading, you can use a strategy that utilizes options only, but creates the same effect as trading the underlying entity. The strategy is called the "synthetic" long or short. Let's talk about the long first.

You create a synthetic long position through the purchase of a call option and the sale of a put option, where the two legs of this position share a common strike price and expiration date. Just as the name implies (long), you must be strongly bullish on the direction you expect the market to head.

An example will help illustrate the concept. You do an analysis of the stock market. Your conclusion is that the Dow is headed much higher—you think nothing can stop it. Instead of taking a position in an underlying futures, like the Standard & Poor's 500, you opt to trade a synthetic long. This means you will buy a call and sell a put. For this example, the S&P 500 is trading at 350. Assume, for simplicity's sake, that the call and the put are in-the-money. It costs you two index points to buy the call, and you are paid two index points for selling the put. Therefore, you break even, less the transaction costs. Some brokers call this a "free" trade.

You decide to hold the synthetic long position until expiration. If prices shoot up as you forecasted, the call gains value and the short expires worthless. You win on both by closing out or exercising the call and collecting the premium on the unexercised put.

What happens if prices drop? In this case, you lose double. Your call expires worthless, causing you to lose the premium you paid, and the trader to whom you sold the short exercises it. You are required to deliver a short futures position that is in-the-money.

The third possibility is the market moves sideways. Here you break even, less transaction costs. You keep the two-point premium you received for selling the put and lose the two-point premium you paid for the call. Keep in mind that the transaction costs are doubled on this strategy, compared with just taking a position in the underlying market. Table 2–4 illustrates the three scenarios described.

T A B L E   2–4

Long Synthetic

S&P Index	P/L Call	P/L Put	Combined	Futures
346	4	2	6	6
354	2	2	4	4
352	0	2	2	2
350	(2)	2	0	0
348	(2)	0	(2)	(2)
346	(2)	(2)	(4)	(4)
344	(2)	(4)	(6)	(6)
342	(2)	(6)	(8)	(8)

This table illustrates how the synthetic options position (row 4) mirrors the performance of a straight position (row 5) in the futures market. Negative numbers are in the parentheses.

Both options do not have to be at the same strike price. If the call is in-the-money and the put is not, you have a deficit since the premium you receive for the put will not cover the call. On the other hand, if the put is in-the-money and the call is not, you'll have a credit. The only problem with this is that the underlying entity will have to move farther in order for your call to start making money.

THE SYNTHETIC SHORT

You can trade a synthetic short just as you could a synthetic long. For the synthetic short, you buy the put and sell the call. Again, you should use the same strike price and expiration date. This strategy mimics taking a straight short position in the stock or futures market.

The same risks mentioned for the long are experienced if the market goes against you. In this case, you are looking for a bearish price move. If prices go up instead of down, your call will be exercised and your put will expire worthless.

As with any investment, you must do your prognostication first. If you call the market correctly, you win the brass ring. If you are wrong, you pay the piper.

Synthetic trades are often used by traders who have a strong conviction about the direction a market or individual entity is headed—and they cannot afford to outright buy the stock or make the margin on the futures. If you find yourself in that position, please don't risk anything you cannot afford to lose.

THE SPECIAL DOUBLE

A unique trade, alluded to earlier, available to options traders is the "special double option," sometimes just referred to as a "double option." This is a combination of a put option and a call option at a fixed strike price. The premium cost of a double is usually close to, but slightly less than, the sum of the premium costs of both a put and a call purchased separately.

The stipulation, which makes it different from the straddle, is that with a double option only one side can be exercised, but both sides can be traded as many times as desired. All the trading techniques discussed earlier with regard to puts and calls work equally well with either side or both sides of the double.

The advantage, of course, is that it frees the holder from dependence on market direction predictions. The double option, being the combination of both a put and a call, can be profitable regardless of whether the market advances or declines. You may often find it easier to determine that there is going to be some price volatility, rather than the direction the market will go next.

Just think about it for a minute. Let's say you notice your favorite commodity or stock beginning to trade in a tight coil or triangle formation. The coil gets tighter and tighter. You know the commodity is poised for a breakout.

But will it go up or down? You're not really sure. All you know is it can't trade much longer without committing itself. This is the opportunity you're looking for to trade the double. When the breakout occurs, you close out the losing side and ride the winning side to the next point of resistance. Then you decide if you wish to take your profits or not.

Like any trading strategy, it is not without risks. For example, the breakout could fizzle. You could end up holding worthless options. The key is correctly evaluating the degree of volatility, which I'll discuss in a later chapter.

BUTTERFLY SPREAD

Here's a useful strategy for quiet markets. All it takes is patience and low price volatility. The term "butterfly" comes from the schematic that is often drawn to illustrate the relationships of the various legs or sides of this spread.

You can execute either a long or short butterfly spread. The long spread is most common, so I'll describe it.

The long butterfly spread entails the purchase of a high strike price and a low strike price option and the writing of two intermediate price options. All the option legs are calls in the long butterfly and puts in the short. For example, you could buy a $7.50 call (high), sell two $7.25 calls (intermediate), and buy a $7.00 call (low). Naturally, multiple contracts could be bought as long as the relationship is kept in balance.

With this strategy, you realize your profit if the market hovers around the short strike price level. Assume that the $7.50 call (out-of-the-money) is acquired for $800, the two $7.25 calls for a total of

$2,300 (at-the-money), and the $7.00 for $1,800 (in-the-money). Combining the purchase prices and the premiums paid, this butterfly costs you $300 plus transaction costs. If everything goes wrong, the options offset each other. The purchased calls are exercised to counterbalance the ones written and exercised. This is your maximum risk.

What happens if the market becomes heated and the price of the underlying entity sinks well below the lower-priced call or rises above the intermediate options ($7.25) triggering their exercise. If it stays in the range of the intermediate options, the low-price option gains 25 cents in value. If the contract is for silver or soybeans containing 5,000 ounces or bushels, this would be a gain of $1,250 less the $300 debit for a $950 net profit times the quantity of spreads being traded. (See Figure 2–4.)

F I G U R E   2–4

Calculating Profit/Loss from Spreads

Strategy:

To profit from a mildly moving market at the least cost.

Assumption:

Trader buys one leg of the spread and sells (writes) the other.

Both options are for the same underlying entity and are both calls or puts.

Calculations:

If premium for the option bought is larger than the premium received for the leg written, the result is a debit in the trader's account.

Futures Example: $.15 premium paid less $.10 premium, received = 5-cent debit.

Stock Example: $4.00 premim paid less $2.00 per share, received = ($2.00)

Debit is then multiplied by the per unit quantity of the underlying entity.

Futures Example: $.05 debit × 1,000 oz. (silver) = ($500.00)

Stock Example: $2.00 debit × 100 shares = ($200.00)

Profit/Loss:

If the underlying entities move in the opposite direction of the options (down for calls or up for puts), all the options expire worthless and the trader loses the amount of the debt.

If the underlying entities move in the direction favorable of options (up for calls or down for puts), the profit generated will be limited by the strike price of options written. When that is hit or exceeded, the holder of the option will exercise it and the trader must exercise his/her option to offset.

SPREAD RISK

Before we leave the subject of combination trades (spreads, straddles, strangles, special doubles), I'd like to alert you to some of the special problem areas you could encounter. First is the problem of getting a good fill on all legs (sides) of the combination trades. Some spreads that are very liquid (heavily traded) have floor brokers that make a market in them and your order can be placed as a spread. Your broker places the spread by ordering the price spread desired.

Each leg of less commonly traded spreads has to be acquired separately, sometimes in different trading pits. In these cases, you have less control over the actual spread between the two options or legs.

If your combination trade includes selling one or both legs, you could get exercised on the option(s) you've written. This would usually only happen if the sell side was losing money for you. Your strategy would be disrupted and the remaining leg may or may not be profitable.

There is always the possibility that both sides of the combination could be unprofitable. These combination trades depend on both sides of the trade reacting somewhat in unison to the same news, so that they close or widen together. There is always the possibility that unexpected or unknown factors will influence one leg and not the other.

When you decide to unwind (offset) your combo, one leg may be more liquid than the other. This particularly happens with out-of-the-money options close to expiration. If this occurs, you may only be able to offset one side. This would leave you holding the other leg as a net long or short position, which may or may not cost you money. Lack of liquidity can be more of a problem with individual stocks than with commodities. With either, it is a very serious consideration that can be checked out in advance.

WRITING OPTIONS

Keep in mind that I'm discussing American-style options, which can be exercised (therefore assigned) at any time, as opposed to European, which can only be exercised at a predetermined period, usually just before expiration. This is a very important distinction, as we begin the discussion of writing options.

If you choose to write options, the fear or risk is that you will be unexpectedly assigned a losing position. And as I said previously, the holder of the option is not likely to exercise until the option is in-the-money. This means that the opposite positions assigned to the writer or grantor are losing money.

Explaining the assignment of an option to a customer by a broker can be a tricky situation. The reason is each exchange has its own rules governing assignments. The Securities and Exchange Commission (SEC) for stocks and the Commodity Futures Trading Commission (CFTC) for futures require the assignment process to be fair and unbiased.

Exchanges use various random selection processes to assign positions, after they learn from their clearing operation, usually an independent corporation, how many option buyers have given notice to exercise options. It can take a day or two before all this works its way back to the writer of the option. Your broker calls: "By the way, you're on margin call. Two of your options were exercised the day before yesterday. Please wire $2,000 by today's close of business."

That's the risk—what's the reward? The benefits of options writing depend on motivation. For some, it can just be a way of increasing the return from a stock. An investor, for example, owns a block of a well-known stock on which options are traded. The return in dividends and appreciation amounts to 10–12 percent per year. To try to push this up to 15 percent, the investor decides to write some covered options on half the block of stock. The expectation is that the options will never be exercised.

A covered option writer is one who owns the underlying entity (stocks or futures contracts) upon which he or she is offering options. An uncovered option writer writes options without owning the underlying stock or commodity. If exercised on, the writer must pay the brokerage firm for the underlying entity which is purchased on the open market. It's important that you have an understanding of the different rules and regulations for trading stock and futures options. They are discussed in the next two chapters.

As with all option traders, the writer must have an opinion of market direction. Will it be up, down, or sideways? Will it be a strong or mild move? The advantage that writers have is that they can win without being accurate.

For example, say your analysis indicates the price of a stock or a commodity is going higher. You decide to sell a put. You would win (collect the premium) if the market trades sideways, if it goes higher, or even if it only goes down (against your position) a little. Part of this calculation depends on which strike price you selected. Option writers only have to be near right to earn the premium, while the investors who buy puts or calls have to be very, very right for the underlying entities to move far enough to pay back the premium, transaction costs, and then a little more to prompt them to exercise the option. Sellers win with singles; buyers need homers, or at least triples!

But profit potential is fixed! All the seller can hope for is the premium, less the transaction costs. The exercise price of the option is the ceiling of profitability for option sellers of covered options. Beyond this point, the option buyers will exercise their right to the underlying entity, taking it away from the seller. The underlying entity could be a stock, foreign currency, debt security, or commodity for covered option. For uncovered, it could additionally include a wide variety of indexes. As mentioned earlier, option writers are uncovered or naked when they write options without owning the underlying interest. Indexes are all uncovered because there is no practical way to own an entire index in the cash market.

Maximum Gain, Loss, and Breakeven Chart:

	Basic Call*		Basic Put*	
	Long	Short	Long	Short
Maximum gain	Infinite	Premium	SP – premium	Premium
Maximum loss	Premium	Infinite	Premium	SP – premium
Breakeven	SP + premium		SP – premium	

*SP = strike price

An example or two might bring this concept into sharper focus. I'll start with a covered call stock option. ABC, Inc.'s stock is currently trading in the $30-to-$35-a-share range. You write a 40 call and pocket the $4 (100 shares × $4 = $400 less $40 transaction costs = $360) premium. Before the expiration date arrives, the stock increases to $47 and the buyer exercises. Your total return would be the $40 the buyer pays you for the stock, the $4 premium (less commission and fees), and any dividends that may have been paid to shareholders of record prior to the date the option is exercised. You do not get any of the gain in stock value between $40 and $47, and you no longer own the shares.

What if you made this same investment, but you didn't own the stock? You would sell a $40 option on ABC, Inc., naked. When the option is exercised at $47 a share, that's what you have to pay to get the shares to meet your obligation to the buyer of the option. If the stock jumped to $50 or $60 a share on news of a merger before exercising, you have to meet the market.

With covered options, you have a predetermined price for the underlying security or commodity. With uncovered options, your cost is whatever the market happens to be when you are assigned (exercised upon) a position or when you go to the market to cover the option you wrote. It is for this reason that the risk of selling uncovered options is considered greater and characterized as unlimited.

Many security firms require new customers to maintain 100 percent of the money needed to fulfill the obligation to deliver the stock for uncovered options. This is called a "cash-secured option." These funds are usually held in an interest-bearing account. In the futures markets, you must maintain the proper amount of margin money to be able to immediately accept the assignment of a futures contract(s). The premium you receive from the buyer can be used to meet all or part of the margin requirement. The futures-type settlement procedure allows you to withdraw excess equity if the funds (premiums, etc.) exceed the margin requirements. Interest isn't usually paid on futures margin, unless it 's held in the form of a Treasury bill, which usually requires a $10,000 minimum.

Two related questions come to mind. First, if you write uncovered options, as most speculators do, how do you protect yours from danger? And which options are most attractive?

Taking the latter question first, the most desirable options to write are usually the most risky. Specifically, these are options that are at-the-money or very-near-in-the-money with a short amount of time remaining to expiration. These options normally have the highest premium because they are the ones buyers think will be in-the-money soonest. They are actively being bid for, which means they are liquid (plenty of buyers).

In response to the first question, you protect yourself by not being greedy or getting married to your decisions. Greed sometimes causes traders to hold positions longer than they should. Traders who become enamored with their analysis have difficulty admitting they are wrong. Thus they hold onto their positions too long.

You protect yourself by dumping losing trades fast. In the industry, we say: "Cut your losers short; let your winners run." If you have written (sold) an option, you offset by buying an option. You were paid a premium to write the option. When you offset, you pay the premium back to another trader who is writing (underwriting) the option. When you are offsetting a losing position, the premium you pay to "buy your option back" is normally more than what you were paid—thus you lose.

It's common for traders to use what are known as stop-loss orders. These are orders to buy a position(s) opposite theirs when a certain price is reached in the market. That certain price is the amount they are willing to lose on a given trade.

Theoretically, this can be done with options. The problem a trader can experience is that many of the options markets are illiquid. In the markets, a stop-loss order becomes the bid or ask price, depending on the direction of the market. The stop loss can be prematurely hit. Another problem is that stop-loss orders are "not-held" orders for floor brokers on most exchanges. In the very liquid markets, stop-loss orders can be very effective.

Another alternative is to transform a naked option to a covered option. You do this by buying the entity that underlies and offsets the option. For a stock option, you'd buy or sell 100 shares of the stock. With an option-on-futures, you acquire a long or short position in the underlying futures contract. Be sure to offset properly. If you wrote a put, you need to buy a long futures contract to offset.

Many traders rely heavily on their brokers to protect them from losses related to adverse market moves. Brokers follow the markets minute to minute. They keep an eye on volume and price activity. They have a good feel for whether a good fill can be obtained during certain trading sessions. The traders may give the brokers some limited discretion on the time or the price at which to place the offsetting order. This permits the brokers to act fast, if necessary. Promptly exiting losing positions is a characteristic that distinguishes the novices from experienced traders.

As mentioned earlier, the most sought after options are often at-the-money or in-the-money, with little time left to expiration. But they are not the only game in town. There are a lot of out-of-the-money and even deep-out-of-the-money options bought. And someone has to be writing these options. A futures option that is out-of-the-money with 60 to 90 days to expiration may only sell for a penny or two.

Are they good deals for writers? A copper call, a dime out of the market with 3 months to expiration, may pay the writer a premium of $0.017, or $425 (25,000 lb. × $0.017). A lot can happen in the 4 months to expiration. The big question is whether copper will gain the 10 cents needed to make it worth exercising by the buyers. Every 1-cent move in copper amounts to a $250 change (25,000 lb./100 cents) in the intrinsic value of an in-the-money option. Therefore, copper must really move about 12 cents before the option writer begins to lose.

If copper begins to move, gaining a nickel . . . then up 7 cents, what do you do when it's at-the-money? When do you close your position? The answer depends on your trading system, which I'll discuss later. My point is that you cannot always sell an option, collect the premium, and passively wait for the option to expire worthless.

Although it is true most options expire worthless, you never know for sure how many times they changed hands before expiration. Many novice traders don't think about this. They believe they can sell options and simply wait for them to erode into nothingness. This can be a very dangerous and costly point of view.

One last thought. The serious sellers of options trade multiple lots. When they sell, they do it in bundles of 10, 20, 50, or 100 options at a time. A hundred lot of the copper options described earlier brings premiums of $4,250. The margin required would be $9,000 to $10,000. If the trader kept a $10,000 Treasury bill on deposit and rolled the 3-month position over 4 times in the course of a year, the trader would more than double his or her margin money each year—plus the T-bill would be collecting interest.

Am I recommending a selling strategy to new traders? No, for these reasons. Before attempting this strategy you need to have a "good feel" for the markets you are going to trade. Good feel means knowing how the market is likely to respond to market news and noise. For example, what is the historical range of price volatility? If you study the long-term price volatility charts, you get an insight in the maximum-size price move you could expect. Some traders think about this as the maximum risk when evaluating the different strike prices. This concept is so important that a later chapter has been devoted to it.

Next, for your protection, you must have a solid working relationship with a successful option broker. It takes time to find and nurture such a partnership. Also necessary are reliable trading and price forecasting systems. All of these subjects will also be covered in detail in later chapters.

My point is—because of the serious risk and predefined reward—that writing options is a step you should not undertake until you have gained experience. Even then, covered options or spreads, which include writing one leg, should be experimented with first.

Now, let's discuss the other major type of options trader.

HEDGING WITH OPTIONS

As mentioned at the beginning of this chapter, hedgers wish to transfer the risk of owning something (cash, a commodity, stock shares) to someone who wants to speculate on the price of that something. Hedgers straddle the price fence, thus the term.

For example, an Iowa farmer normally produces 150 bushels of corn per acre on his best 500 acres, or 75,000 bushels. By June, his crop looks good and the futures price for December corn is at $3.75. The farmer wants to lock in some of his production at this price. So he decides to hedge 50,000 bushels, or 10 contracts of 5,000 bushels each.

Since he will have physical possession of the corn in December, after harvest, he is already long. To hedge or become neutral he must short the market or sell 10 December corn at-the-market puts. This locks in the $3.75 (less his commission cost for the option contract) per bushel price on 50,000 bushels.

If the price of corn falls, the farmer's puts gain in value to compensate for the loss in the cash side of the hedge. The key is knowing the delta or hedge ratio of the strike price selected so the farmer will be 100 percent hedged or neutral pricewise. If the strike price had a hedge ratio of 0.50, the farmer would buy 20 puts instead of 10.

This example could have used copper, silver, foreign currencies, or stocks. With stock portfolios, the hedger uses the stock indexes that most closely reflect the composition of the portfolio. For broad-based portfolios, consider the S&P 500 or the New York Composite Index, the Value Line for portfolios of small stocks, the

Over-the-Counter Index for nonlisted stocks, or the Major Market for portfolios similar to the Dow Jones Industrial Average.

To protect against a price rally, the hedger buys calls. They gain in value as prices of the underlying entities gain. If your company had to pay a Japanese manufacturer in yen for a piece of equipment 60 days or 6 months from now and you were afraid the yen was going to rise against the dollar, you might buy yen calls to "freeze" the current conversion factor. Or if your business uses large quantities of a commodity, like copper, silver, corn, soybeans, cattle, money, foreign currencies, etc., and you wanted to protect yourself against a price increase, you could buy puts at your price, convert them to futures contracts, and take delivery when the contracts matured. You would be removing the price risk factor from these transactions.

Hedging with options is like buying price insurance for the commodity you own or must acquire in the future. (See Figure 2–5.)

MARKET OPINION

All of the strategies discussed have one key element in common—the trader must have an opinion about market direction. I discuss how one arrives at an opinion in Chapter 4.

F I G U R E   2–5

Soybean Hedge

	Cash	Option
May Prices	$5.75	Buy Nov. $6.00 put at $0.25/bu premium
Sell Cash at Harvest (Nov.) Price:	$5.00	Sell Nov. $6.00 put at $1.00 in-the-money
Strategy: Sell (buy a put) Profit:	$0.75	
Cash Received	$5.00	
Option Premium Received	$0.75	
Total	$5.75	

CHAPTER 3

A Few More Uses for Options

"What can't be cured must be insured."

Oliver Herford (1863–1935)
American Writer

Key Concepts

- Using Options to Close Open Stock or Futures Positions
- Options as Insurance against Negative Price Swings
- Scale Trading the Futures Market with Options

A common use of options, in the course of regular stock or futures trading, is to close positions. This trading technique can generate additional income and afford some risk control or management in certain instances.

When you can sell an option, you contractually agree to deliver the underlying entity (most commonly 100 shares of stock or a futures contract) to the buyer (long) of the option. If you already have the stock or the futures contract(s) in your trading account, you are writing a covered option, as described in the preceding chapter.

You can use the same approach to close positions. I once had a client who traded this way. She sold options just above major resistance points. For example, she owned IBM stock. In one period of its trading history, IBM had difficulty penetrating the $100-per-share level. It would touch this level and then retreat. Obviously, there were a lot of traders with sell stops in place at that level, perhaps even some computerized trading systems.

My client's strategy was to sell covered calls at $105, which was the target price for the stock at that time, a very good price considering where she had purchased it. If that strike price was hit—she delivered the stock. If not, she'd keep the premium and wait for her target price to be hit. A simple, but effective, use of options. She also bought options on stocks she thought were going to split, since there is usually a lot of activity (volatility) in a stock at that time. Again, a good use of options.

OPTIONS AS INSURANCE

Using options as insurance is common to both stock and futures trading. If you fear your stock will be under some severe pressure, due perhaps to negative earning reports or a pending law suit, but you don't want to sell the stock or take the financial loss, you buy enough puts to cover your position. When the stock price drops and levels off, you sell your puts. This profit cushions the loss the stock experiences. But like all insurance, it is only another expense if it is not needed.

This type of insurance is even available to mutual fund investors. The Chicago Board of Options Exchange now has options on two of Lipper Analytical/Solomon Brothers mutual fund indexes: the Growth Fund Index and the Income Fund Index. Prior to these options, it was impossible to cover an entire mutual fund investment.

Naturally, these options are not for insurance only. If you want to speculate on the direction in which mutual funds are headed, you can. Or you can use them as a hedge and lock in your profit after a fund has had a major move to the upside.

The Growth Fund Index and the Income Fund Index each contain the 30 largest mutual funds having the same investment objective. Annually, these indexes are reviewed and adjusted to ensure they continue to include the 30 largest funds. In addition, they are rebalanced quarterly to be dollar-weighted and adjusted for dividends and other distributions. LEAPS, or long-term options a year or more in length, are also available. But these options are European style, which means no exercise before expiration. The strike prices are set in 5-point increments and are calculated in 100-point multiples.

For example, if you purchased a 200 strike price put, it would protect $20,000 of your mutual fund(s) from a major downdraft. The cost would naturally vary depending on the market psychology at the time.

OPTIONS AND FUTURES STRATEGIES

Using options as insurance—to get into and out of positions—is as common in futures trading as it is with stocks. For example, when I was a branch office manager for Securities Corporation of Iowa, the commodity brokers there utilized a trading system called "scale trading." Now the scale trading purist never advocated the use of options, but some of our traders and brokers adapted them nevertheless because they were so useful. Scale trading is one of the few futures trading systems I've seen that has merit.

First, let me give you a quick overview of scale trading. It attempts to extract a modest profit, approximately 30 percent per year, from the futures markets by following its most basic laws— those of supply and demand. If a commodity is plentiful, the price decreases. If scarce, it increases. Or if a commodity is heavily in demand, its price will reflect positively. If there is no demand, the price wanes. These forces can work in tandem or in opposition. Plentiful supply will temper price rises in conditions of vigorous demand. The converse is equally true. As is the fact that if a commodity has strong demand and is in short supply, prices will soar. Prices plunge when supplies are in excess and demand is weak.

A corollary to the laws of supply and demand states: The surest cure for low prices is low prices. This simply means that once prices get to a low enough level, people find uses for that commodity. For example, when the price of wheat gets below a certain level, cattle feeders substitute it for corn. Another example: Gasohol and sweeteners were developed when corn prices languished at low levels. If something is cheap enough, long enough, someone will begin buying it, driving the price higher. Thus a low price cures low prices.

Scale traders attempt to make a profit from these "natural" laws. They buy futures contracts of commodities trading in the lower 25th percentile of their 10-year trading range, called the "buying zone." They hold the contracts until prices turn around. The difference between scale traders and other speculators is that scale traders set the exit price at the time they buy, or open a position, at a predetermined profit margin. Thus you get the term "scale trading"—a commodity is bought on a scale as it drops and sold on a scale as it rises.

For example, you study the 10-year price chart for corn. The high is $4.00 per bushel and the low $2.00, or a $2.00 range. The lower 25 percent would then be $2.50 or the starting range for buying corn on a scale.

Let us imagine over the next 2 months that corn drifts steadily lower, eventually coming to rest in the $2.00 area. Using a scale of buying every 4 cents down and selling every 6 cents up, you would start buying at $2.50, then $2.46, $2.42, etc.—until corn bottoms at, say, $2.02. Every penny loss for a corn contract is $50, or 5,000 bushels times $0.01. Therefore, the oldest position is losing 48 cents, or $2,400. The total loss of all positions is $13,200. Review the corn scale in Figure 3–1. Additionally, you would be required to maintain margin money on all 13 positions. Margin is always subject to change with-

F I G U R E   3–1

Corn Scale

	Buy Prices ($ per bu.)	# of Contracts	# of Losing Positions	Losses per Contract
1.	$2.50	1	1	$2400.00
2.	$2.46	2	2	$2200.00
3.	$2.42	3	3	$2000.00
4.	$2.38	4	4	$1800.00
5.	$2.34	5	5	$1600.00
6.	$2.30	6	6	$1400.00
7.	$2.26	7	7	$1200.00
8.	$2.22	8	8	$1000.00
9.	$2.18	9	9	$ 800.00
10.	$2.14	10	10	$ 600.00
11.	$2.10	11	11	$ 400.00
12.	$2.06	12	12	$ 200.00
13.	$2.02	13	(at the money)	$ 0.00

$13,200.00

out notice, depending on the volatility of the market, but let's call it $400 per contract for this example. That comes to an additional $5,200, for a total of a negative $18,400.

How long you'd be required to hold these positions is unknown, but eventually corn would move higher—it always does. A true scale trader must believe that what goes down will go up. The question is, can you wait long enough and do you have the financial staying power to hold 13 losing positions for 3, 6, or 9 months?

Corn could start moving up immediately, or you would have to wait indefinitely. When I was scale trading back at SCI, some markets, like cocoa, regularly moved up and down rather smoothly. Cocoa "behaved" well because the governments in Africa that controlled much of the world's supply held it off the market when prices got too low.

When corn begins to move higher, you start lifting positions. Your scale calls for a 6-cent profit on each position. Therefore, when corn hits $2.08, you offset your most recent position ($2.02) and pocket $300 ($0.06 × 5,000 bushels). If corn moves steadily up, as it did down, when it reaches $2.56, you have sold all 13 positions with a $300 profit on each for a total of $3,900, or 21 percent of the total invested ($3,900/$18,400).

That's how it works theoretically. But rarely does a commodity move so smoothly from one trading range to another. The more normal pattern is for a commodity to stair-step up and down. It may move a nickel or a dime higher or lower, only to make a 66 percent retracement. Then it lurches again, followed by another recovery. Eventually it reaches an equilibrium.

This jerkiness can be either good news or bad news for the scale trader. For example, the intermittent bull rallies during a bear market move can generate additional profits. This occurs when commodity prices move high enough to trigger a trade, 6 cents in the example above. Then the market moves low enough, 4 cents, to reenter the trade. This is called an oscillating profit opportunity. The good news: If you get enough of these, you move your ROI into the 30 percent range, which is the goal.

The bad news about an oscillating market is that it can prolong price movements, which means some of the early positions can expire and must be rolled over into more distant contracts. For example, a June corn futures contract must be rolled to a September one. The

September contract is trading at a higher price and the scale can be put out of balance. The true scale trader patiently waits out these adjustments. Others close out these positions at a loss, especially if they have been in the scale a while and have made a decent profit already.

A major oscillation could really be a change in the market trend. In other words, the market changes direction and carries the price out of the scale trader's trading range. In this case, you simply take what profits you've earned to date and look for another market to scale.

In my opinion, it is this discipline—strictly trading the scale's entry and exit points and walking away from any market that is not within a trading range—that accounts for the success traders have had with the system. There is a lesson here for you no matter what system you decide to trade.

HOW DO SCALE TRADERS USE OPTIONS?

The scale trader is expected to hold onto long positions at all costs. But margin money can be a problem, particularly if a commodity drops unexpectedly fast or farther than expected—below all-time lows. Don't forget futures contracts can have limit-down days when they fall like lead balls in the ocean and no trading takes place. If you are long 5 or 10 positions and your commodity limits for 2 or 3 days, you may be in margin trouble. With the limit-down days, you can't even offset positions at a loss. You must wait until trading resumes.

But what if you had sold some calls at your exit points? As the market plunges, they would become more valuable. When trading resumed, you could exercise those positions and get out of your longs. This is not what a scale trader is supposed to do, but if you can't meet margin, you'd be tickled pink to do it.

Or you could sell calls against your long positions, covered calls. Then you could use the proceeds to buy puts as insurance. Traders call these "free" puts because you're not investing any new money into the market.

If the market goes up, you deliver your long positions against the calls you sold at your scale's exit positions. You're even, except for the cost of the puts, which is usually less than your profit target. So you're still a net winner.

If the market goes south, you exercise your puts to offset your losing long positions. Or you can sell the puts at a profit and use the proceeds to feed your margin calls as you hold your long positions until the market turns around.

What if the scale trader is not sure when to enter a market? The commodity may be in the buying zone, but acting bullish. You think that at any minute it could rally and move out of the buying zone. What do you do?

One strategy is to sell calls as an entry strategy. Let's say you sell a call and get 6 cents for it. If the market bounces up and you get hit, you're long or you've begun your scale. But you still have the 6 cents, which means that if you got this at the beginning of the scale, $2.50, your actual position is $2.44.

One last word about scale trading. It is more complicated than it may appear in this discussion. You must have access to good fundamental information about the worldwide supply and demand situation—production, weather, political conditions, shipping, the list goes on and on. Then you need a broker that really understands the system, particularly when to begin a scale, which market is ready to scale, and how to roll over positions when that becomes necessary. I've seen the system work, and I've also seen traders get into it underfinanced only to exit broke.

These are just a few ways to use options in conjunction with your stock or futures trading system. They definitely have a place in both.

CHAPTER 4

Forecasting Stock and Futures Price Trends

Question: "What are the desirable qualifications for any young man who wishes to become a politician?"

Mr. Churchill: "It is the ability to foretell what is going to happen tomorrow, next week, next month, and next year. And to have the ability afterwards to explain why it didn't happen."

Sir Winston Churchill
(Adler's Churchill's Wit, *1965)*

Key Concepts

- Basics of Fundamental Analysis
- Understanding the Five Primary Technical Approaches to Market Analysis—Bar Chart Analysis, Trend Following, Structural, Character-of-Market, and Other Approaches
- Pros and Cons
- Which Works Best? When?

Option traders must have strong opinions to deliberately invest in what they think stock or commodity prices will be tomorrow, next week, next month, and next year. When they are wrong, they must have the ability to reevaluate and correct their analysis—and then get right back in the market! This chapter and the next one deal with the most common approaches taken by traders to solidify their thinking.

First, I'll discuss how to forecast the price trends of the under-lying entities. These trends dictate the price movements of the options on those entities. In the next chapter, I tackle the concept of volatil-ity and theoretical option price modeling compared with the actual bid-ask prices generated from the trading floors.

There are two basic schools of thought on price forecasting—fundamental and technical analysis. One utilizes the right side of the brain; the other the left. One is more intuitive; the other more ratio-nal. One looks for the reasons why prices will change; the other directs its attention on past patterns in its search for future direction.

Fundamental and technical analyses are used to anticipate price trends for both stocks and commodities. Most techniques were ini-tially developed for stocks, since they have been trading longer on organized exchanges in the United States (the Philadelphia Stock Exchange was founded in 1791; the Chicago Board of Trade in 1848). Analysts choose their approach based on their education, back-ground, experience, past success, and temperament. Some combine the techniques. If you plan to trade options in an organized fashion, you will probably find yourself primarily in one camp or the other.

FUNDAMENTAL ANALYSIS

I'll begin with fundamental analysis because most people find it is easier to understand. It is basically common sense. All you do is study the factors that can change price and anticipate their impact. For commodities, when supplies become scarce, prices rise. When plentiful, prices recede.

The analyst builds a model that represents the supply-demand equation. Think of it as a tug of war with supply on one side and demand on the other. The bigger and heavier the supply side gets, the easier it is to pull prices down. If demand grows bigger and faster than supply, it will pull prices higher. Here's the basic supply-demand equation for a commodity:

Existing Stocks plus Production minus Usage = Supply

The supply is then evaluated as being adequate (sideways price trend), inadequate (bullish market), or overabundant (bearish pro-jection).

Fundamental analysts get gray hair from unexpected events. They project a drought that will reduce production of a grain, only to see rainfall and record-setting harvests. Or an Iraq invades a Kuwait—then plays "cat'n'mouse" with UN arms inspectors—upsetting the world supply of crude oil. Strikes, government actions, changes in consumer preferences, new inventions, etc.—all these can ruin an otherwise brilliant fundamental analysis.

Another perennial problem for the fundamental analyst is gathering and analyzing all the possible variables. Just getting all the information (for commodities—weather, political activity, government programs, usage, currency, yields, input prices/supply, demand, etc.) from all over the world can be prohibitive for most individuals. This is exaggerated when other countries are uncooperative or deliberately misleading, to improve their country's situation.

You must also consider the fact that fundamental analysis is usually of a long-term nature. When you begin to access information about the supply side of the equation, like the planting intentions of farmers or the impact of long-term interest rates on farmers, the implications of this information on future prices will not be known immediately. And a lot of other events will occur in the intervening time.

Fundamental analysis of the stock market is somewhat different. The supply of shares of a stock doesn't change (a new issue of stock that dilutes holdings of current stock being an exception) from one season to the next. And when the supply changes, as in a split, the price usually reflects the change and is normally made on an undiluted basis.

Demand for stocks is more closely related to the general health of the economy. Fundamental stock analysts often study business cycles. As Charles Dow taught—when the tide rises, all the ships in the harbor rise. As the Dow average rises, so do "all" stocks. By knowing when the economy is about to change direction (heat up or cool off), the analysts can anticipate general demand or lack of interest.

One of the key factors is the indexes prepared by the Bureau of Economic Analysis of the Department of Commerce. The indexes were chosen from all the data the department gathers based on the following six characteristics: economic significance, statistical adequacy, consistency of timing at business cycle peaks and troughs, conformity to business expansions and contractions, smoothness, and prompt availability. Analysis indicated that combining certain of these individual indicators into indexes was more reliable than depending on them individually. Below are the results of this research.

Index of Leading Indicators

Average workweek
Average weekly jobless claims
New orders for consumer goods
Vendor performance
Contracts/orders for new plants/equipment
Building permits
Change in unfilled durable goods orders
Sensitive material prices
Stock prices
Money supply (M2 in 1982 dollars)
Consumer expectations

Index of Coincidental Indicators

Employees on nonagricultural payrolls
Industrial production
Personal income less transfer payments in 1982 dollars
Manufacturing and trade sales in 1982 dollars

Index of Lagging Indicators

Percent change of labor costs per unit of manufacturing output
Ratio of manufacturing and trade inventories to sales in 1982 dollars
Average duration of unemployment
Ratio of consumer installment credit to personal income
Commercial and industrial loans outstanding in 1982 dollars
Average prime rate change by ranks
Change in consumer price index for services (smoothed)

The indexes are distinguished by their timing characteristics. The Index of Leading Indicators historically reaches its cyclical peaks and troughs before the general business cycle does. The Index of

Coincidental Indicators makes its turnaround at roughly the same time as the business cycle does. The statistical indicators that make up the Index of Lagging Indicators typically hit their highs and lows after business cycles have turned.

If you study the composition of each index, you can see that the statistics cover just about every aspect of the economy. You find labor, production, management, consumers, financial, and even stock market statistics included. Most importantly, these specific statistics were selected because they so closely tracked the ups and downs of business cycles.

The rule of thumb is that turning points in the economy are signaled by three consecutive monthly readings in the same direction. For example, if the Index of Leading Indicators records a +0.2 percent in July, a +0.4 percent in August, and a +0.3 percent in September, one would be theoretically headed out of a recession, if the country is in one. The opposite is equally true—three negative monthly readings in a row indicate a recession on the horizon. These indexes are not 100 percent accurate. Most economists are happy with a 70 to 80 percent reliability. Much also depends on momentum. Large, consistent changes are more meaningful than weak, inconsistent ones.

STOCK MARKET REACTION

These indexes seldom have immediate market impact when released. The reason is simply that the components are known to the market before they are compiled into indexes. Serious analysts watch the individual statistics as they are released. It's easy from them to predict the movement of the indexes based on changes in the components.

Like everything else associated with options trading, there are occasional surprises. These usually involve unexpected changes in degree, rather than direction, of the indexes. When these occur, the reaction of the market is the same as it would be toward any other macroeconomic indicator. For example, higher or more positive numbers in the Index of Leading Indicators foster the expectation that economic expansion is just around the corner. The bulls in the market begin buying, while the bond traders prepare to short their markets.

Analysts must put the numbers into the perspective of the business cycle. Sometimes strong leading indicator numbers can be received negatively. This occurs when they are released late in the business cycle. At these times, they are seen as harbingers of growing inflation. They are telling economists that the economy may be about to expand too fast. Inflation may be out of control.

Also, there are times when these three indexes give contradicting signals. One or two may indicate the economy is expanding, while the third suggests the opposite. These types of circumstances are common to traders. The situation tells traders to stand aside until all three synchronize. As with any market prognostication system, you must know when to bet the factory and when to stay on the sidelines. Always remember the indexes are called "indicators," which denotes that they are not expected to give you the absolute direction of each upcoming move in the economy.

Once the overall direction of the trend of the stock market—as reflected by the business cycle—is clear, the fundamental stock analysts evaluate those stocks they feel will most benefit from the trend. It is common at this stage to look at the subindexes, such as the Dow Jones Industrials, to see which subcategory is displaying the most strength. Earnings of stock classified by industry (pharmaceuticals, insurance, publishing, etc.) can be checked. There are 97 industry groups in common use.

The last step is selecting individual stocks and making specific recommendations. At this stage, analysts study earnings per share, talk with executives, project growth of the industry, and calculate liquidation value. There are literally hundreds of ways to fundamentally pick the stock(s) with the most growth potential. Most individual investors must rely heavily on their brokerage firm to do the exhaustive task of research and analysis.

TECHNICAL ANALYSIS

Technical analysis (TA) is completely different. Pure technicians totally disregard everything but the signals generated by their technical system. The theory is that every fundamental factor affecting supply and demand reflects in the price. Therefore, if you study the stream of prices flowing from the trading floors and the related statistics, such as volume, open interest, and momentum, you can uncover trends and anticipate future price objectives.

Don't let technical advocates mislead you. Technical analysis tracks the past; it does not foretell the future. Technicians rely on history repeating itself. There is one thing TA does very well, and that is quantify market sentiment. It provides an excellent technique for counting all the votes of everyone trading.

Think about the stock and futures markets as public polls. The brokers are the pollsters. They go out and survey the general trading public. "Where do you think the price of the S&P is headed?" "In your opinion, is the price of IBM too high or too low?" The answers to all the questions are communicated to the trading floors, where they are processed. The results are published daily in various financial papers in the form of price gains or losses.

Keep in mind that the people who answer these surveys are very sincere about their answers—they have backed their opinions with money. Collectively, they represent all the thoughts, research, and emotions surrounding the entire financial community. The people polled may have been influenced by the media, their personal financial situation, a computer program, or what has happened and is happening in the world.

The voting results show up on price charts, momentum oscillators, and other technical analysis tools. These tools are useful ways of measuring the temperature or tone of the markets. New behavior patterns manifest themselves in new price chart formations. Your task, as a technical analyst, is to make sense of them. When you start this process, you join the rest of us in attempting to guess at what will happen tomorrow.

Technical analysis is often referred to as a self-fulfilling prophecy. This evolves from the fact that its use is so widely accepted, particularly in the futures markets. For example, one of the simplest techniques is drawing trend lines on price charts. The most common example of this is the up or down trend line. Every technical trader is taught that when a major trend line is broken, it's time to reverse one's position. And just like lemmings, when prices break the trend, "all" the traders reverse their positions—fulfilling the prophecy. For this reason alone, anyone trading options should know the basics of chart analysis, if nothing more than to anticipate what other traders can be expected to do.

One of the most compelling reasons for having faith in technical analysis is the enormous amount of trading equity controlled by technical traders. The reason for the popularity of TA is simply the enormous amount of information required to trade successfully using fundamental analysis. The natural consequence of acquiring gigantic amounts of data is not having the time and resources to do the analysis. Therefore, technical analysis is more manageable and controllable, especially with the use of computers. It often gives traders a false sense of security.

Another little quirk of fundamental analysis is that it is not self-correcting. For example, if a trend line is broken, or a support-resistance level penetrated, the technical trader views this as a reason to reverse or abandon a position. But what does the fundamental analyst think if he or she believed corn was a good buy at $2.00 per bushel and it drops to $1.50? Is it a better buy? When does one reverse one's position? When do you "know" you're wrong?

Before I go too far, I want to stress the fact I am not belittling fundamental analysis in any way. There are major grain companies and well-known professional traders (CTAs) who use it as their primary analytical technique. And most stock traders scoff at technical analysis. It is also very common for all types of traders to use fundamental analysis to determine the basic trend or the overall direction of the market. Then, technical analysis is used to select specific exit and entry points.

Another problem with fundamental analysis is unknown factors. We have seen many unexpected political moves in recent years—the fall of the Berlin Wall, the Iraqi invasion of Kuwait, to name just two—that have sent the market reeling uncontrollably. How do fundamental traders handle these? It is hoped they do it with protective stops.

Technical traders will see the market breaking a trend line or generating an exit signal of some sort, and react. Technicians have "reasons" to reverse or stand aside (not trade) the market(s).

Please don't get the impression technical analysis is a panacea. Traders lose money just as easily and just as fast with technical analysis as they do using fundamental analysis. The big difference is there are technical analysis systems that are manageable by small traders.

One thing I have learned over many years of options trading is that it is not the system alone that generates success for traders. If there were one or two perfect systems, we'd all know about them and they probably wouldn't work any more. But there are hundreds and thousands of systems.

To make a system work, you must combine it with sound money management techniques and stick with it. The most valuable aspect of a trading system may be that it "forces" you to make a trading decision by flashing a signal. By constantly fine-tuning a system, you can make it work for you. Your system's and your money management rules must make you a disciplined trader— one who can survive the inevitable losses you will sustain. In other words, no analytical system—fundamental or technical—can fore- tell the future!

TYPES OF TECHNICAL ANALYSIS

I divide technical analysis into five basic types—bar charts of prices, trend following, structural, character-of-market, and "other." Price bar charting is probably the most common for commodities. Daily, weekly, and monthly prices are charted and the formations analyzed. An example of trend-following technical analysis is moving aver- ages, a technique that combines a series of prices and smooths them mathematically. Structural analysis assumes the market moves in established, recognizable patterns—like seasonal, cyclical, or wave patterns. Probably the most sophisticated is character-of-market. This type of analysis attempts to exactly measure the "quality" of a price movement and then takes a position opposite the momentum, anticipating where the market is headed next.

As you might guess, the last category, "other," is a collection of all types of analysis that are not easily classified. This could be any- thing from something as simple as trading contrary opinion to insights gained from studying the phases of the moon. I'll now pro- vide an overview of each type.

Bar Charts

Since there are only 40 or so commodities (with 4 to 12 monthly con- tracts) actively traded compared with thousands of stocks, it is eas- ier for a publisher to create a chart service for them. Additionally, most stock traders hold their positions longer, which reduces the need for and value of a daily charting service. Nonetheless, charts of stock prices can easily be located. Options charts can be found in some advisory newsletters devoted to options traders and can be generated by some software programs. (Refer to Appendixes 2 and 4 for specific names, addresses, and telephone numbers.)

Current charts are so important to traders that most electronic price quotation services provide them—even for on-line data. Brokers commonly follow electronic charts showing every price tick within seconds of when the trade occurs.

The basic element of the price bar chart is the price tick (Figure 4–1). The vertical line represents the unit's (day, week, month) trading range. The top is the high, and the bottom is the low. The short horizontal line on the left denotes the opening prices, and the one on the right (as you face it) is the close. Not all charts include the left or opening price tick. Most price charts show only a closing tick. The vertical axis of the chart represents price and the horizontal time.

Daily, weekly, and monthly bar charts are depicted similarly with the high, low, and close represented by a vertical line above

F I G U R E   4–1

Price Ticks

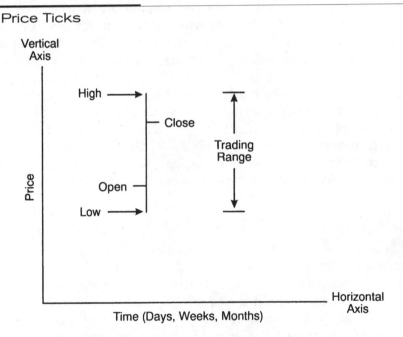

The price tick is the basic element of commodity price charts. The vertical line denotes the price range for the period (day, week, month). The left horizontal tick mark indicates the open, and the right horizontal tick the close.

the date. Nontrading days, weekends, and holidays are not depicted, so there is no break in the pattern.

Long-term charts of commodities, covering months and years of price activity, periodically show price distortions up or down. Careful analysis often reveals these are places where the contract rolled over from one contract month to the next. For example, the June silver contract expires and the chart service continues the chart using the December contract. In this case, there could be a price gap or distortion between the two contract months. The reason is simply that as the June silver contract approaches expiration or delivery, trading heats up, driving prices a little higher or lower. The far-out contract, December, trades with less volatility, and prices remain steady. My point is that the rollover from one contract to another can sometimes cause a distortion, depending on the factors influencing the different delivery months. These longer-term charts are often referred to as "continuation" charts.

As a rule, the longer term the chart, the smoother the pricing pattern. This means that the price patterns on weekly charts are usually smoother than those on daily charts, and that monthly charts are smoother than weekly ones. Also, the longer the term of the chart, the more reliable the pattern is considered. For example, an area on a monthly chart that has given support to prices (I'll be discussing specific price patterns shortly) would be expected to be stronger (or more reliable) than the same formation on a weekly or daily chart. This is generally true for all formations or chart signals.

Across the bottom of most daily charting services (see Figure 4–2), you'll find a record of the trading volume and open interest for that day. Volume indicates the total number of contracts traded. Open interest measures the number of contracts held at the conclusion of a trading session. These are very important figures because they tell you the degree of activity of traders, or the momentum of the market you're studying. I'll be discussing these in more detail as we discuss specific price patterns.

Chart Formations

You could fill a good-size bookcase with full-length texts discussing interpretations of bar charts for stocks and commodities. Several of the

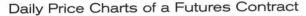

FIGURE 4-2

Daily Price Charts of a Futures Contract

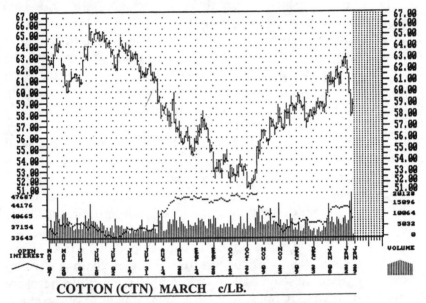

COTTON (CTN) MARCH   c/LB.

Here is the daily price activity of a futures contract. Weekends and other
nontrading days are omitted to provide a smoother picture. Daily volume
and open interest are plotted across the bottom of the chart.

better ones have been cited in Appendix 2. What I am doing is high-
lighting some of the more common formations to give you an insight
into how a trading system based on bar chart analysis might work.

The most common and significant formation is the trend line.
It can be either an uptrend or downtrend line. Trend lines require at
least two points to touch the line and are considered more reliable
the closer they are to 45 degrees in incline or decline. A parallel line
is sometimes drawn to indicate a channel or trading range. Uptrend
lines are drawn below the price ticks and downtrend above. The
higher degree or steeper the trend lines are, the likelier they will be
broken and the trend will reverse (see Figure 4–3).

All technical analysis is built on the premise that history repeats
itself. When a trend line is established, technical traders expect it to
continue. When it gets too steep, they expect the trend line to be broken. When a trend line is broken, they expect the trend to reverse.

FIGURE 4-3

Uptrend Lines

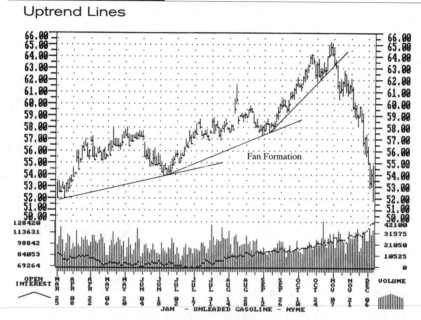

Fan Formation

Trend lines indicate the overall market sentiment. It could be bullish, bearish, or neutral. The steeper the trend lines become, the more likely a trend reversal will occur.

Chart courtesy of "Pocket Charts"

This mind-set is part of what makes technical analysis work—what makes it a self-fulfilling prophecy.

Technical analysis also works because it is based on sound psychological principles, the most basic of which is herd psychology. When the members of the trading herd see a trend line, they trade it until it ends. When it ends, they look for the next trend or formation and trade it.

No one knows what the future will bring. It is a total unknown. Therefore, people (traders) who risk their money on anticipating where commodity prices will go next are uncertain, uncomfortable. When people are uneasy, they stick together. This basic need of humans is responsible for history repeating itself and the meaningfulness of chart formation. Just like anything else in life, none of the formations are anywhere near 100 percent reliable.

Support and resistance areas are very common and significant. There are places where downtrending markets come to rest repeatedly or uptrending markets tend to stall repeatedly (Figure 4–4). For example, a market rallies and then falls back to a previous price level, from which a second rally is launched, only to decline to the same price level again. This is a support area. The opposite, where a market rallies time and again, only to stall out at a certain price level, is a resistance area.

Areas of support and resistance are "zones of error." Thus, once a support area is penetrated, it becomes a resistance area because there are traders who bought at the support level and still hold their losing positions. They'd be happy at this point to get out at breakeven. These traders offset their losing positions, which props the opposite side of the market.

Closely related to support and resistance formations are multiple tops and bottoms. Take the double top as an example. A market

F I G U R E   4–4

Areas of Support and Resistance

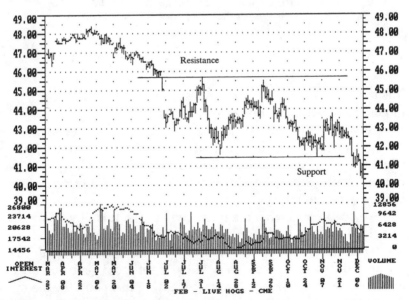

Areas of resistance slow or stop uptrends; support levels retard or halt downtrends.

rallies, reaches an area of resistance, stalls, and begins to decline. After a short retracement, it rallies again, only to stall at the exact same resistance area. The chartist would call this a double top or "M" formation and consider it a sign of weakness. It would be a signal to short the market. The risk, of course, is that a triple top could develop. The trader who shorted the double top would be whipsawed by the market. Whipsawing is when you reverse your position only to have the market turn against you. You lose twice, once on your original position and again on your reversal. Multiple bottoms are the opposite of multiple tops.

More complex versions of the multiple tops and bottoms are the head and shoulders and inverted head and shoulders formations. They resemble the silhouette of a person (Figure 4–5) and are very important for several reasons. First, they are considered very reliable by professional chartists. Furthermore, they foretell a major reversal in trend. Last of all, they can predict the length of the next move or the reversal.

F I G U R E   4–5

Inverted Head and Shoulder Formation

WASENDORF GRAIN INDEX, 1977 = 100

Head and shoulder formations resemble a silouette of a person. Note how this one signals astute technicians of a major upcoming price move.

A head and shoulders formation usually consists of the following five key phases:

1. Left shoulder.
2. Head.
3. Right shoulder.
4. Neckline penetration.
5. Prices often retrace and bounce off the neckline before
   heading toward their first price objective. This provides a
   second opportunity to short the market.

The first price objective is measured from the top of the head to
the neckline and projected downward from the point where the neck
was broken. A second, or even third, objective of the same distances can
be projected, depending on the velocity of the market as it reaches the
first objective. Inverted head and shoulders work just the opposite.

Another formation that is considered very reliable by chartists
is the rounded bottom. It is a long, drawn-out formation that can
take months to mature. It is often called a "saucer" bottom because
of its shape. When the saucer bottom matures, it signals a long-term
uptrend. The reason for this is that while the rounded bottom is
maturing, the price of the commodity is relatively low. It is at the
bottom of its price range for a long time. While this occurs, it is not
uncommon for new uses to develop that increase the long-term
demand. For example, while corn was making a rounded bottom in
1986–87, new uses for it (as noted earlier), as gasohol and corn sweeteners, were developed. When it finally began a price rise, it had a
stronger demand base, and a long-term bull move developed.

A lot of other bar chart patterns are used by chartists, such as
triangles, boxes, and key reversal days, but I can't cover them all. If
you're serious about trading, you need to spend some time studying all the formations—so you'll at least know what other traders
are likely to do.

Trend-Following Techniques

The second category I use to classify technical analysis techniques is
trend following. I'd like to quickly review just two basic approaches.
The first is moving averages.

Moving averages can be calculated for any period of time—3, 5, 7, 30, 200 days, whatever. Once calculated, they can be charted (Figure 4–6). To prepare the moving average, you can use the high, low, open, or close. The latter is most common. Naturally, you must consistently use the same one. Here's an example of how a 3-day moving average is constructed:

Day	Closing Price	3-Day Total	3-Day Moving Average
1.	2.40		
2.	2.42		
3.	2.45	7.27	2.42
4.	2.50	7.37	2.46
5.	2.53	7.48	2.49
6.	2.55	7.58	2.53
7.	2.53	7.61	2.54
8.	2.49	7.57	2.52
9.	2.45	7.47	2.49

You can see that this procedure smooths out the price volatility. The moving average lags the current price activity on both the way up and the way down.

Some traders have developed trading systems by using combinations of moving averages. For example, you could use a set of moving averages of different lengths, a 5- and a 10-day. The slowest moving average, the 10-day in this case, gives the long-term trend. The faster moving average, the 5-day, provides the buy-sell signals. The rules are simple:

1. Buy when the faster average crosses above the slower.
2. Sell when it crosses below.
3. Offset long positions when the daily price closes below either moving average.
4. Offset shorts when the daily close is above either moving average.

The challenge is to uncover the set of moving averages that works best for the stocks or commodities you wish to trade.

Another sample of a mechanical trend-following system is the point-and-figure chart. It is based on the theory that long periods of

F I G U R E   4–6

Price and Moving Average Chart

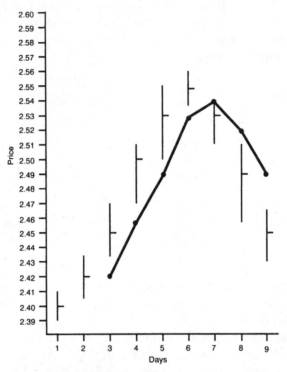

Note how this 3-day moving average lags the daily prices. Moving averages are used to smooth price action as a means of clarifying trends.

price consolidation are required to generate a significant price move. This system totally disregards time and concentrates solely on price movement.

Chartists use "X"s to denote upticks and "O"s for downticks. Therefore, much interday information is required and a scale must be set. Computer optimization programs are available to select scales. The only rule of thumb is that the scale must be larger than the minimum price fluctuation of the contract being charted. A popular scale for commodities is 1-cent value to each block with a 3-cent reversal. This simply means that an X or an O is put on the chart for each 1-cent move up or down. Once three consecutive Xs or Os are recorded against the current trend, the trend is considered reversed and positions are offset or reversed (Figure 4–7). With this system, 45-degree angle trend lines and channels are considered to be even more significant than on bar charts. Other formations, like double bottoms and triangles, are also used in the analysis.

STRUCTURAL ANALYSIS

Technical analysts who use this type of analysis believe in the "natural law." The sun rises in the east each day. The seasons rotate from spring to summer to fall to winter. Patterns and respective cycles can be found everywhere. Physical sciences are based on this fact. Why not stock and commodity prices? The structural analysts search for these patterns. Discussed below are a few of the more prominent schools of thought.

FIGURE   4–7

Point-and-Figure Chart

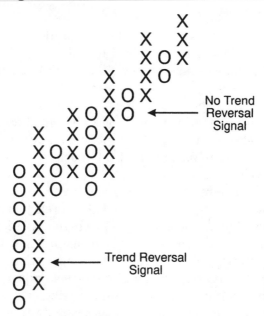

The "X"s denote positive price moves and the "O"s negative. These charts can be used as a mechanical trading system within themselves.

SEASONALS

Seasonal patterns are the easiest to understand, especially if you consider the agricultural commodities. The seasonal price trends these commodities undergo reflect the regular annual changes that take place in their supply-demand equation.

Think about a typical crop year. First, we anticipate planting intentions. Then we worry and fret through the weather markets of the growing season. Finally we usually have an abundance of supply at harvest time.

Studies have shown, for example, that 70 percent of all seasonal tops occur between April and July for soybeans. These studies have also shown that 80 percent of the time soybean prices bottom between August and November. How can you use this information? If you know the seasonal patterns for the stock or commodity contracts you trade, you can use these patterns to confirm signals you get from other methods of analysis. Some traders will not trade against reliable seasonal patterns without having a very strong reason for doing so.

Cycles

Cycles are similar to seasonal patterns, but they can be longer than 12 months or extremely short in duration. Cycles are built on the observed phenomenon that events have a tendency to repeat themselves at reasonably regular intervals. The business cycle described earlier would be a good example.

As mentioned earlier, much of a person's life is governed by repeatable patterns or cycles. Since people made and drive the markets, it seems fair to assume that the markets would also possess definable cycles.

Cycles measure the time between highs and lows (peaks and troughs). By knowing the time span between each high and low and the previous high and low, you are in a better position to anticipate the next high or low. Time is usually measured in calendar days, as opposed to trading days. Calendar days are used for the simple reason that people and nature do not take weekends off—money continues to change hands, as do events affecting cycles.

A long-term cycle generally lasts a year or more; an intermediate one less than a year; a short-term one a few weeks or days. As a rule, allow approximately 10 percent leeway in the length of a cycle when establishing your expectation for the next top or bottom.

Elliot Wave

The Elliott wave, also known as the wave principle, states that social and market behavior trends and reverses in recognizable patterns. These patterns or waves recur. Prices unfold in five waves of crowd psychology when moving in the direction (up or down) of the primary trend. Then they move against the trend in three waves. The wave pattern reflects life's starts, stops, false starts, reversals, etc. Progress is made in a jerky, sawtooth pattern, rather than a smooth uptrend or downtrend.

Isolating the exact position of the current price activity within the wave patterns allows the trader to profit by anticipating the market's next move. Once your trading becomes synchronized with the wave pattern, you can successfully ride the economic waves of the markets. This technique is most commonly used to analyze the stock indexes and the Dow.

Gann Numbers

The best known of all structural analysts is probably W. D. Gann. His 1942 book, *How to Make Profits Trading Commodities*, was the first important treatise on this subject. He believed that precise mathematical patterns govern everything, particularly the stock and commodity markets. More importantly, he believed these patterns could be uncovered and exploited. Since his price predictions became legends in his own time and he claimed to have made millions in the market, his followers, even to this day, assume he discovered many of them.

Integral to his trading system are Fibonacci numbers—again a throwback to natural law—and angles of price trend movement. His numbering system uncannily alerts traders to highs, lows, support-resistance areas, and reversal points. His work should be studied by every serious trader.

Again, this is not a definitive list of structural analytical systems. I just want, in this text, to give you an overview. To continue in this vein, I'll outline a few points about character-of-market–type systems.

Character-of-Market Analysis

I consider character-of-market analysis to be very sophisticated because it attempts to measure the quality of a price movement, and then take a position that may be opposite that of the current trend. This differs from the other types of analyses discussed so far, which try to spot existing trends or certain formations that are reliable harbingers of future price activity.

With character-of-market analysis, the technician seeks what are known as "overbought" or "oversold" conditions. If the market is found to be overbought, it is sold. If it is found to be oversold, it is bought. This approach is a 180-degree turn from what I talked about with the trend-following approaches. With them, a strong market is bought and a weak one is sold.

The psychology behind character-of-market analysis is simply that when markets become too top heavy, they fall. Or when everyone gives up on prices ever rising again, they will. It is a contrarian approach to technical analysis.

The trick is determining when a market is overbought or oversold. Here are some of the better-known approaches.

Oscillators

Oscillators are concerned with price changes over a period of time. Simple oscillators utilize the difference between two moving averages. The departure between them indicates overbought and oversold conditions.

More complicated ones use the difference between daily prices. It can be the settlement, high-low, or opening price. Take a simple 5-day settlement price oscillator as an example. It is computed by subtracting the settlement price of the fourth previous trading day from the current settlement price. If the settlement price has risen, you get a positive remainder. A negative remainder occurs if the price has fallen. If the remainder is exceptionally high or low, the analysts will consider the market overbought or oversold and take the opposite position. Analysts using oscillators usually use more complicated ones than this simple example. But the basic concept is the same.

Relative Strength Indexes

These are indexes that attempt to quantify the momentum of the market. When certain price levels are reached, a market is said to be overbought or oversold and, therefore, due to reverse its trend.

This approach works well in choppy, zigzagging markets. If you use it in a long-trending market, you may get burned badly. You need many tools on your technical analysis workbench. These indexes are excellent ones to have when the market appears to be confused or erratic—but never rely on just one approach to analyzing all markets and all the varying market conditions.

Other Approaches

I can't spend a great deal of time on "other approaches" to technical analysis for two reasons. First, there are simply too many of them. Second, most of them are too complex to be easily described. The best I can do is make you aware of a few of the possibilities.

Astrological Analysis—Astrology is the study of the aspects and positions of planets and other heavenly bodies to predict future human events. There are a number of books, and even a newsletter, devoted to price analysis based on astrology. Lunar phases, planets, and even sunspots have also been used.

Correlation Analysis—This is the study of the relationship of a specific stock or commodity with a related or nonrelated entity in the hope that a correlation between the two (or more) will indicate future trends. A simple example is the correlation between gold and paper currencies, gold and the U.S. budget deficit, or a particular stock and the advance-decline numbers.

Fibonacci Numerical Progression Analysis—The classic name for this mathematical phenomenon is the golden ratio. It was originally recognized by the ancient Greek mathematicians and popularized in the thirteenth century by Leonardo of Pisa, known as Fibonacci. The numerical series is created first by adding 1 and 1. The rest of the numbers in the series are found by summing the last two terms to get the next number in the sequence: 1, 1, 2, 3, 5, 8, 13, 21, 34, 55, 89, 134, etc. Using these numbers in various calculations as divisors or multipliers generates the golden ratio, which is 1.618 or 0.618. This ratio was called golden by the ancients because it appears widely in nature (branching of trees, flowers, seashells, etc.) and is appealing to the eye. It was used in such varying structures as the Egyptian pyramids and the Sistine Chapel.

If the golden ratio exists in nature, is used commonly by people, and creates a pleasing design or formation, then, technical analysts postulate, it should appear in price charts as well. When it does, they give special emphasis to price advances or retracements that reflect the golden ratio. These technicians look for moves up or down of 62 percent or angles of this degree between critical points. No one is absolutely sure why this ratio is important, but in many cases it appears to work and generates meaningful projections. W. D. Gann, as mentioned earlier, made extensive use of the golden ratio in his work.

CONTRARY OPINION

One last type of analysis you should be aware of is sometimes referred to as "contrary opinion." It simply states that the masses are usually wrong. If the crowd or herd thinks a specific market is headed higher, it is probably ready to make a move lower.

The stock, futures, and options markets are very fast-moving, very emotional markets. Traders are called on to make split-second decisions, often involving large sums of money. Additionally, these decisions are not necessarily based in cold, logical analysis of facts. Often, as you probably gathered from this discussion of fundamental and technical analysis, the decisions are based more on faith than anything else. This is only natural since you're dealing in future events.

For this reason, you need to guard against joining a stampede of bulls or bears. You must walk the line between following a selling trend and getting caught up in a groundswell of emotion that keeps you in a market too long—namely, beyond the point where it turns dramatically against you, generating substantial losses. Contrary analysts believe that by the time the herd reacts, it is too late and things are about to change. Contrarians can base their decisions on either fundamental or technical analysis.

COMBINED APPROACHES

Using one type of analysis, even if one approach is technical and the other fundamental, does not preclude using others, even several others. The most common example is the analyst who uses a fundamental system to determine the long-term trend and one or more technical systems for the short term. The fundamentals tell you why the market is moving; technicals signal when. This is sometimes referred to as the "market view approach."

I specifically said "one or more" technical systems for the short term because certain systems work better in certain types of markets and for certain stocks or commodities. For example, at times the markets in general, or a specific stock or futures contract, may have the tendency to trend up and down for long periods—3 or 4 months or longer at a stretch. When this occurs, trend-following systems (moving averages, etc.) work well.

But what happens when the markets get very choppy or trendless? Trend-following systems suffer severe whipsawing in nontrending markets. You probably want to use a character-of-market type of system in these very erratic types of markets.

My point is that no single trading approach will be successful in all market conditions. If the activity of a market doesn't suit your system(s), you should change markets or stand aside until more favorable conditions return.

APPLYING YOUR ANALYSIS

The purpose of your technical and/or fundamental analysis is to solidify your opinion of where, and how strongly, the market is headed. You select your option trading strategy based on your degree of bullishness or bearishness. There is at least one other alternative, which is the subject of the next chapter.

CHAPTER 5

Option Price Models and Volatility—The Most Important Consideration for Serious Option Traders

"Most of the change we think we see in life is due to truths being in and out of favour."

Robert Frost's Black Cottage

Key Concepts
- Using Theoretical Option Price Models
- Determining If an Option is Overvalued, Undervalued, or Fairly Valued
- Random Walk Price Theory
- Calculating and Quantifying Stock and Commodity Price Volatility
- Analyzing Historical, Seasonal, and Implicit Volatility
- Combining Volatility Analysis with Fundamental and Technical Forecasts

In our survey of technical analysis, I deliberately neglected to mention one very important tool—theoretical option price modeling. My reason is simply that a full chapter needs to be devoted to the subject.

Option prices, as determined by open outcry on the various stock and commodity trading floors, are either fairly priced, overvalued, or undervalued. They must be one of these three. Therefore, if you could compare what the price "should be," given the current facts, with what it is in the pits, you could come to a logical opinion regarding the most advantageous trading strategy.

For example, if your analysis indicates that the option being considered is fairly priced and that it will remain that way until expiration, you might consider writing the option. On the other hand, if all indications are that it is grossly undervalued and poised to make a major move higher, you'd look for the best bullish strategy. The opposite, or a bearish strategy, would be called for if you are confident the option is currently overvalued by the marketplace.

Another important use of price modeling is strike price selection. Your fundamental or technical analysis alerts you to a major bullish or bearish move—but which of the five or six strike prices available offers the most profit potential? Price modeling provides a rational, quantitative answer. In other words, price modeling can be used in combination with other types of analysis, as well as a stand-alone trading technique.

Part and parcel of this concept of overvalue and undervalue is price volatility. How widely will prices swing in the next 12 months, quarter, month, week, or day or two? Once you know the volatility and the option's price relationship to the current market, you're in great shape to select a strategy and/or strike price(s).

The best way to begin is to introduce you to the Black-Scholes option pricing model. It is not the only model available, but it is the best known and most widely used. The model was developed in the 1960s for stocks and modified in the mid-1970s for options-on-futures by Myron Scholes and Fischer Black. Both men are legends in academic circles and the investment community.

The formula for the Black-Scholes model is complex, and beyond the scope of this book. Option traders use computerized versions requiring the input of four variables. The software program solves for the fifth. These programs are widely available and even affordable (see Appendix 4), and virtually all brokers who trade options have access to them.

The variables used are:

1. Time to expiration of the option
2. Price of the underlying entity (stock or futures contract)
3. Exercise or strike price of the option

4. Carrying charges (interest rate, dividends for stocks)

5. Volatility of the price of the underlying entity

If you were evaluating an option, it is easy to see that the first four items are given. You can count the days to expiration. The current price of the stock or commodity comes from the paper, quote screen, or your broker. You select the strike price(s) you want to evaluate, and the carrying charges are assigned or, with the case of dividends, are announced in the financial press or can easily be estimated. With interest rates, it is usually recommended to use a rate that is closest to the time to expiration. For example, if there are 3 months to expiration, you use the 3-month Treasury bill rate.

The last item, volatility, causes all the grief. It is the one input item that can be wrong, resulting in an incorrect value. This could be very costly. For this reason, I'll discuss volatility in some detail. To understand price volatility and the Black-Scholes model, I must back up a little and talk about how prices change and the assumptions on which the model is based.

RANDOM WALK

An enormous amount of study has been done over the years by some very intelligent and well-equipped economic scientists in hopes of discovering what causes prices to change. To the best of my knowledge, no one has discovered the secret.

As mentioned in the last chapter, traders and analysts use fundamental and technical analyses in their quest for discernible price trends, but these analyses are by no means 100 percent reliable. Most market participants would be ecstatic with a reliability factor anywhere near 100 percent. Computers offer the promise of becoming the market tamer. Their capability of making thousands of calculations per second, of easily handling the most complex equations, and of manipulating millions of bits of data in econometric models is the hope of economists who dream of predicting tomorrow's prices today.

To date, there is no definitive price modeling methodology for futures, stocks, or options. But isn't this what excites us about these markets? Try as we do, our success rate is limited. The better we do, the harder we try—and the more frustrated we become. Sounds a lot like golf, doesn't it?

Not being able to predict prices or price trends into the future with any certainty leads to the conclusion that price movements are random in nature. An additional conclusion is that prices cannot be artificially manipulated, at least for extended periods of time. In other words, the few attempts at cornering a market failed. The most recent example was the Hunt brothers' assault of the silver market in 1979–80. They bought, bought, bought silver until it approached $50 per ounce—only to watch it, and the family's wealth, plunge. Cornered markets are the rare exception that proves the rule.

More importantly, there are some price movements that appear to occur regularly and can be predicted mathematically. This characteristic of pricing allows us to determine the probability of future price action. The key word is "probability." It will never, in my estimation, be possible to unequivocally forecast prices. If traders or investors knew, as opposed to thought they knew, what tomorrow's prices were definitely going to be, they'd discount the impact of inflation and everyone would agree on a price. There would be no need for exchanges.

The factor known as inflation is an important component of price determination because it modifies the concept of normal distribution of prices. If prices were absolutely distributed normally, they would be evenly scattered on either side of the mean, or average price. Inflation challenges this logic. A normal distribution curve is symmetrical, allowing for values of from below minus to infinity. Therefore, is it logical to expect negative futures or stock prices?

A symmetrical pattern also suggests there is an equal chance for an asset to go down as to go up in price. Inflation tends to improve the chances for price increases the farther we project into the future. This, of course, does not eliminate the strong possibility of a commodity or a stock losing money or going down in price. Nor does it remove the possibility of the percentage of growth to be lower than the inflation rate, resulting in a de facto loss of value. Nonetheless, the price of virtually all commodities and stocks gains over the long term.

What price analysts do is skew the results a little toward the positive side of the mean as they build their equations. This is accomplished through the use of a lognormal distribution curve. It calculates price changes in percentages, rather than dollars and cents. Prices move up or down as a percentage of the current price. A stock or commodity currently priced at $100, for example, experiences a

5 percent move. That puts it in a range of $95 to $105. At $50, the 5 percent move is $2.50. As the entity approaches zero, the changes become nearly meaningless—just as it is illogical to expect the value of a stock or commodity to go to zero.

At some price, someone will find a use. Corn, for example, during the Great Depression was used as fuel. When stock prices go "too" low, someone buys the stock and liquidates the company. As prices go higher, the dollar increments become greater, even if the percentage remains constant.

At this point, let's summarize the conclusion of this discussion about price changes:

1. Price changes appear to be random, since there are not any analytical tools that are anywhere near 100 percent reliable. At best, fundamental and technical analyses systems are only partially successful.
2. Prices defy artificial manipulation over the long term. Eventually, supply and demand come into balance.
3. Prices move up or down. No one can accurately predict with any certainty which way they will move next.
4. Price changes, in dollar terms, should be skewed toward a positive side of the mean to be realistic using lognormal distribution because of inflation and other upward pressures.

These are the assumptions behind the Black-Scholes commodity and stock option pricing models. Most other theoretical pricing models also use these assumptions.

Now, let's build on these assumptions. Since price changes are random, you need to have an understanding of how prices are distributed randomly. This in turn leads to an appreciation of how to locate the mean (average price) and how to calculate deviations from the mean (volatility). The beginning point is a quick review of the theory of probability.

THE FLIP OF A COIN

If you ever have been forced to take Statistics 101 in high school or college, you probably have been exposed to a discussion of the odds of getting a head or a tail when flipping a coin. Extensive experiments have documented that no matter how many times you flip a coin, there is always a 50-50 chance of getting a head or a tail. If you toss 10 straight heads in a row, you still have a 50-50 chance of getting a head on the eleventh try.

This axiom is the cornerstone for the concept of normal distribution of random events. For example, let's flip a penny 225 times—15 separate series of 15 tosses each. The theoretical results follow:

Series Results	#Heads	#Tails
1	1	14
2	2	13
3	3	12
4	6	9
5	8	7
6	7	8
7	10	5
8	11	4
9	9	6
10	5	10
11	6	9
12	4	11
13	2	13
14	1	14
15	0	15

Each time you repeat this experiment, you may or may not get the exact results shown. But what will happen is that the majority of the time you will get very similar results.

Specifically, you'll most often get the results—assuming the coin is properly balanced—in the middle of the spectrum. It would be very unusual (32,000:1) to get 15 straight heads or tails. Or only one or two tails and the rest heads. Or all heads and no tails. Normal distribution describes what is likely to happen with random events. Charting this on a graph generates the well-known bell-shaped curve.

In addition, it is assumed that you need to use lognormal distribution. This skews the results to the positive side of the mean to account for inflation and the unlikely event of a commodity or stock price going to zero. Making this adjustment to the curve produces a bell-shaped curve slightly leaning to the right, as you face it.

The bell-shaped curve is the distribution curve. The one produced by flipping a coin repeatedly is "normal" or moderate in appearance, but the shape can change. Erratic data produce a flat arc because the data points are scattered. Data representing a slow-moving market result in a curve that is very steep or narrow, since the data points are close together (Figure 5–1).

The key concept is that the flatter the shape of the distribution curve, the more volatile the market being evaluated. What you want to learn to do is plot price changes for various markets to generate distribution curves, so that you can quantify market volatility. Once you do this, you will be able to make statistically valid estimates of what price moves can be expected from these markets.

F I G U R E   5–1

Volatility Curves

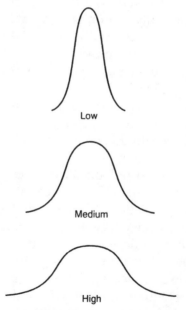

The closer price points are, the lower the volatility. Highly volatile markets are characterized by wide or flat curves, since the data points are spread out.

Underline the words "statistically valid estimates of what can be expected." This is the best you can do when you are attempting to forecast prices. On average, or, normally, you'll be correct. This doesn't mean you'll always be right. You'll still be wrong some of the time.

The hope is that it will put the odds in your favor, giving you an edge. It should provide an excellent insight in market volatility, allowing you to answer questions like:

"How much of a price move can you expect?"

"What are the chances of this market making a limit move?"

"Can you expect a certain futures contract or stock to make enough of a move during the life of the option you plan to buy to generate a profit from the option?"

This type of analysis provides educated guesses to these questions. It is the basis behind the computer programs that calculate the theoretic value of options, as an example. By running these programs, the software ranks the option alternatives, telling you which ones are underpriced or overpriced. It may even create strategies to take advantage of the given market conditions.

This statistical analysis won't tell you which way a market will move. Don't forget, we are discussing volatility of what we have classified as *randomly* moving prices. You will be able to figure the statistical odds of a market making a $2 move over the next 35 days— but will it be up or down?

That information won't come from this type of analysis. The answer—or more correctly, best estimate—comes from other types of analyses, such as the fundamental or technical analysis already discussed.

Now, let's take a closer look at the concept of the mean and the concept of standard deviation. The mean is simply the average of the price data and is the peak of the curve. The standard deviation describes the arch of the curve. Flat curves represent fast-moving markets; steep curves reflect low volatility.

For purposes of illustration, let's go back to our coin-flipping example. We made 15 series of 15 flips. To find the mean, you multiply the number of either heads or tails (only one side need be calculated) in each flip by the number of that flip and divide the total by the total number of heads (occurrences) recorded.

Flips Times		# Heads
1 × 1	=	1
2 × 2	=	4
3 × 3	=	9
4 × 6	=	24
5 × 8	=	40
6 × 7	=	42
7 × 10	=	70
8 × 11	=	88
9 × 9	=	81
10 × 5	=	50
11 × 6	=	66
12 × 4	=	48
13 × 2	=	26
14 × 1	=	14
15 × 0	=	0
Totals 75		563
563 divided by 75	=	7.5067

The mean or average for 15 coin flips is 7½, which is logical since we expected the normal flip of a coin to be a 50-50 proposition. Therefore, approximately half the data points will fall on each side of the bell curve, if we were to plot the results of all the flips.

Calculating the standard deviation is somewhat more involved, utilizing very complex algebraic formulas beyond the scope of this book. What you do need to know is the meaning of the standard deviation.

In our coin-flipping example, the standard deviation calculates to approximately 3.00. Standard deviation tells you two things about the data, namely how scattered or diverse they are and the probability of any specific outcome. In the case of a futures contract or stock price, this means you can calculate how fast a market is likely to change (volatility) and the chances of a certain price being hit or surpassed.

Please don't worry about not being able to manually calculate the standard deviation. Any trader who uses it regularly nowadays relies on computer programs. The function for figuring it is built into most advanced spreadsheet programs, for example. Naturally, it is part of the software that runs the Black-Scholes formulas.

Actually, what is normally done is to calculate more than one standard deviation. It is common to figure 1, 2, and 3 standard deviations from the mean. These standard deviations are on both sides (plus and minus) of the mean.

They tell what percentage of the occurrences (coin flips, closing prices, rates of return) being evaluated will fall on either side of the mean. Here's a guideline:

- Plus or minus 1 standard deviation includes 68.3 percent, or two-thirds, of all occurrences.
- Plus or minus 2 standard deviations include 95.4 percent, or 19 out of 20 occurrences.
- Plus or minus 3 standard deviations include 99.7 percent, or just about all occurrences.

This simply means that if the result you are expecting, known as occurrences, falls within 1 standard deviation of the mean (average or midpoint of the bell curve), you would have a 2-out-of-3 chance of that happening. Don't forget this is plus or minus 1 standard deviation, which means the result could be positive or negative because you have made the assumption that price movement is random.

For example, you want to know the probability of a certain futures contract making a limit move or the probability of a stock price doubling. By running a standard deviation of the price activity, you could calculate the probability of that happening.

But there is some "Kentucky windage" involved. How much data do you use to calculate the standard deviation? Do you use only last week's, month's, or quarter's prices? What is the life of that particular contract or stock? Are all the data on record? This is a judgment call on your part based on experience.

Consider taking two or three snapshots of the volatility using current, near-term, and longer-term data. Is the volatility increasing or decreasing? This provides a feel for the price activity so you can make your judgment.

Does this analysis guarantee the results? No, definitely not. Even at the third standard deviation level there is 0.3 percent chance that the next occurrence (settlement price, etc.) will be outside the range.

To get back to our coin-flipping example, the mean is 7.50 and the standard deviation is 3.00. Therefore, the first standard deviation ranges from plus or minus 3.00 of 7.50 or from plus 4.5 to minus

4.5. This would be a line (look back to our table of results) between flips 4 and 5 and flips 10 and 11. Therefore, all the combinations of coin flips from 5 through 10 would have a two-thirds probability of happening. If we were to flip a balanced coin from now until eternity, the law of probability states that 68.3 percent (approximately two-thirds) of the results would fall within the above-stated range.

Now, I'll discuss an actively trading market. What I will be doing is taking a snapshot of the current price activity and making some projections. In doing this, three things can go awry. First, even if I use the third standard deviation, there is a 1-in-369 chance that the next data point will be outside the parameters.

Second, I could anticipate a positive move and actually experience a negative one. In other words, my analysis of the direction of the next move could be incorrect. Third, I could be using an incorrect volatility value.

The volatility value changes whenever a new data point is added. Now the change may be virtually immeasurable if the new data point falls near the mean, but it occurs nonetheless. It is for this reason you should think of price modeling as a snapshot or single frame of the price discovery process with an emphasis on the word "process." For price discovery is indeed an ongoing process.

The first step in analysis of a trading market is to establish the mean. In most option price models, such as Black-Scholes, the current price is the mean. These pricing models assume the expected return from the underlying position (a futures contract or 100 shares of a stock) is zero. The apogee of the bell curve is the current price. To adjust for the expected return, carrying costs and dividends are added. For futures, there are no carrying costs of consequence, because the interest on the "small" margins, combined with the short period of time these positions are usually held, is not significant. And futures do not generate dividends.

For stocks, it is a different story. At least half of the cost of a stock purchase must be anted up when a position is opened on margin, and interest must be paid on the other half. Additionally, dividends may be paid during the life of the option on the underlying stock. Therefore, the carrying cost and dividends are added to the current price.

Once you have the mean, you can proceed to calculate the standard deviation or volatility. This volatility figure represents 1 standard deviation of price change at the end of 1 year. For example, say gold is trading at $400 per ounce and has a volatility value of 20 percent. Over a year, you would project its trading range as plus or minus 20 percent, or from $320 to $480 per ounce, approximately 68 percent of the time. Calculating the second standard deviation (20 percent × 2 = 40 percent) projects that 95 percent of the data price points will range between $240 and $560 per ounce. The third standard deviation forecasts a range of $160 to $640 per ounce, which should account for 99.7 percent of prices "discovered" by open outcry in the pits over the year in question.

In most cases, using calculations based on a year isn't much help. Few options are liquid enough to be traded over a full year, and even fewer option traders wish to hold positions that long. Sometimes even a month or a week seems like an eternity for in-the-money options. Therefore, you must convert the volatility to shorter periods of time. The Black-Scholes model handles this by dividing the annual volatility by the square root of the number of trading periods in that year.

Here's an example of adjusting the yearly volatility rate to a daily rate. The usual figure used for the number of trading days (exclusive of weekends and holidays) is 256. This number is used because it is very close to the average in any given year and has an even square root, namely 16. Therefore, if the annual volatility rate is 20 percent, the daily rate would be 1.25 percent (20 ÷ 16). This means the $400-per-ounce gold contract would have a daily volatility of $5 or a projected daily range of $395 to $405 per ounce. Of course, this is for the first standard deviation, or 2 out of 3 days, or 68 percent of the time. At the second standard deviation level, you could expect a $10-per-ounce swing up or down 1 out of 20 days. A $30-per-ounce move could be expected 1 out of 369 days, which would be the third standard deviation.

How do you project this probability out for a week? Stocks and commodities trade 52 weeks a year. The square root of 52 is 7.211. The annual volatility of 20 percent divided by 7.211 is 2.77 percent. This figure times $400 per ounce equals $11.08, which is the maximum weekly price change you would expect two-thirds of the time. Only 5 percent of the time would you project a change of more than $22.16 on a weekly basis.

From what I've discussed so far, it is clear how you can calculate the anticipated dollars-and-cents price range for a commodity or stock given any time period and volatility rate. For example, what would the price range of a $100-per-unit commodity or stock share be for the next 25 days with an annual volatility rate of 35 percent?

$$
\begin{array}{ll}
365 \text{ day}/25 \text{ day} & = \ 14.6 \text{ trading periods} \\
\text{Square root of } 14.6 & = \ 3.82 \\
35\% \text{ volatility}/3.82 & = \ 9.16\% \\
9.16\% \div \$100 & = \ \$9.16
\end{array}
$$

The price range would be plus or minus $9.16, or from $90.84 to $109.16 for 68 percent of the time.

The beauty of this mathematical exercise is that you can work backward as well as forward. Given the volatility value and the projected price change, how good a probability is it that you'll see the price change you project? Say you need a $30 price change on a $200-per-unit futures contract or stock between now and expiration, just 50 days away, with an annual volatility rate of 20 percent.

$$
\begin{array}{ll}
365/50 & = \ 7.3 \\
\text{Square root of } 7.3 & = \ 2.7 \\
20\% \div 2.7 & = \ 7.4\% \\
7.4\% \text{ of } \$200 & = \ \$14.80
\end{array}
$$

This tells you that the $30 target may be unrealistic from a law-of-probability standpoint. You should expect a $14.80 swing 68 percent of the time. If you divide $30.00 by $14.80, you get 2.02, or approximately 2 standard deviations. Therefore, you have approximately a 1-in-20 chance of a $30 move. Do you offset your position faced with this scenario and salvage what you can, or do you continue to hold? The answer depends on your money management skills and your approach to risk.

WHICH VOLATILITY?

As mentioned previously, the amount of the data used to calculate volatility varies. For example, data for a week or a month or a year can be input in the equation. Also, any particular price, such as the open, high, low, or close, can be used. The standard approach is to use the settlement price, which is the daily price at which the clearinghouses settle all accounts between clearing members for each contract and contract month. The settlement price and the closing price are not always identical.

In other words, some discretion is required when selecting the data. Two traders, one using the settlement prices and the other opening prices, could calculate different volatility factors for the same contract over the same period of time. The same is true for the two other analysts, where one uses the settlement price for the last 5 days and the other uses 200 days of data.

You must also be aware of the fact that intraday price movements or daily price ranges can be greater than 1 standard deviation price change on any given day. Let's say you calculated 1 standard deviation for a certain contract as 50 cents a day. You calculate from settlement price to settlement price. Your calculation of volatility could be right on-the-money, and yet the daily price range could be $1 or more. (As a side note, other methods have been explored to figure volatility besides the standard deviation, but none have caught on.)

Continuing this line of thought, you may calculate the daily volatility of a certain $50 commodity or stock at 1.25 percent or a dollar value of 62½ cents. For the next 5 days, you monitor the following price changes: +0.55, –$1.25, –0.05, +0.25, +0.75. With 1 standard deviation of 62½ cents, the law of probability indicates that 68 percent of the time the price change of the commodity should range between plus 62½ cents and minus 62½ cents. In the above series of price changes, three out of five, or 60 percent, are within the range.

What should an analyst conclude from this information? First, the data only cover 5 days. Therefore, there is no reason to immediately assume the volatility rate is incorrect. Second, the percentage of price changes within 1 standard deviation is close, 60 percent versus 68 percent, to expectations. The one figure that might concern the analyst is $1.25, which is over twice 1 standard deviation. The expectation would be for a price change of this magnitude to occur 1 in 20 times, rather than 1 in 5.

The first thing the analyst would do is look for a reason why one of the price changes ranged into the second standard deviation. This often occurs when a government report exceeds market expectation. For example, let's say the live hog market was trading at $50 per hundred weight. The USDA (United States Department of Agriculture) then released hog numbers 10 or 20 percent below market expectation, driving up prices the $1.25. If this was the case, there would be a good reason for the surprise price change. If no reason for the change can be found, the analyst might consider changing the volatility rate.

The correct nomenclature for the volatility rates calculated using current market prices is implied volatility. It is the rate of price change implied by current trading. If this is what implied volatility is, what kind of volatility have I been previously talking about? Up until now, I have been discussing historical volatility. It provides a consistent view of an option over a period of time. For example, a service providing daily historical volatility charts would use the daily settlement price, or the service might use each Friday's settlement price to produce weekly charts, or the settlement price of the last trading day of the month for a monthly chart. Refer to Figure 5–2.

The value of the historical charts is they can give you the overview or the long-term perspective. Let's say you calculate, using the Black-Scholes model, that you need a volatility factor of 30 in order to have a shot at making money or breaking even on an option you are holding. Perusing the historical volatility charts for the underlying commodity or stock reveals that volatility has never exceeded 25. Is there any reason to expect a new high to be made? If not, it may be time to offset your position to salvage as much time value as possible.

SEASONAL VOLATILITY

Volatility is the quality of being volatile, which means "changing readily and repeatedly," "changeable," "fickle," or "transient." From my experience, this definition suits the grain markets to a tee at certain times of the year. It also holds true for several other market groups, such as petroleum and softs, that have pronounced seasonal patterns.

The rationale for this behavior is simple. The more widely scattered prices for a commodity, let's say corn, the higher the volatility. Think back to the bell curve. If prices are scattered in a broad pattern around the mean—as opposed to a tight pattern or narrow price range—the curve will be flatter. The flatter the curve, the higher the standard deviations.

Broad price ranges in the corn market are the result of uncertainty regarding the supply-demand situation at key times of the year. The first spurt of volatility occurs when the USDA releases its initial estimate of the planting intentions of farmers. How many acres will be planted? Factored into this figure will be any federal acreage reduction or set-aside programs.

F I G U R E   5–2a

Soybeans (CBT) Volatilities

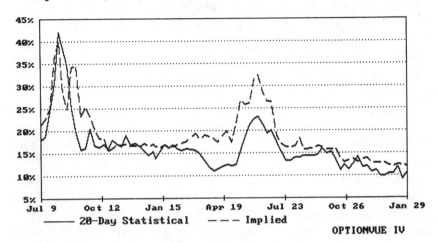

OPTIONVUE IV

F I G U R E   5–2b

International Business Machines Volatilities

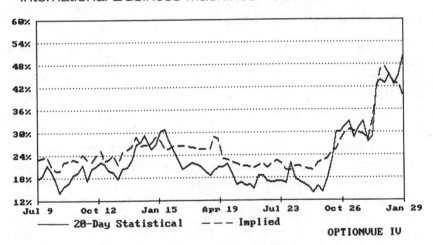

OPTIONVUE IV

   Once the crop is in the ground, a series of weather-generated volatility rallies usually occur. Spring rains bring the risk of delayed planting. But if it doesn't rain, there is the fear that germination rates will be low and that the chemicals (primarily fertilizers, herbicides, and pesticides) won't be activated. After the corn is as high as an elephant's eye, the farmers begin to worry about a drought or pollination problems. Then the worrywarts struggle through a wet fall that could hamper harvest, early frosts that may reduce yields, and wet fields which could keep the combines in the equipment sheds.

So far, I've only discussed the supply side of the supply-demand equation. The demand leg has its own seasonality, primarily tied to livestock numbers, prices, and demand. If ethanol ever gets wide acceptance, corn demand may be influenced by the seasonality of gasoline production and demand.

Price rallies naturally stimulate volatility rallies, if prices are widely scattered. It is not uncommon to see the volatility jump by 3 percent, or even as much as 10 to 15 percent, in a very short period of time. The obvious result of these swings is rapidly accelerating prices for calls, particularly strike prices in-the-money or near-in-the-money. They are being influenced by both the direction of the market, which is increasing in intrinsic value, and the improved volatility, which makes it more attractive to traders.

An informative exercise is to repeatedly run the Black-Scholes model changing only the volatility rate. Keep all the other variables—namely, strike price, time to expiration, underlying futures contract price, and interest rate—the same. This provides an excellent insight into the impact of volatility. I recommend at first varying the volatility by 5 points each time. Then duplicate the drill at just 1-point intervals.

When attempting to trade seasonal volatility, keep in mind that the influences on one contract month may not apply to another. For example, a contract that extends through the July and August drought months, such as the November contract, should be assigned a higher volatility rate than the May contract that doesn't. You might give November a volatility factor of 5 percent more, compared with May.

When you begin to assign or adjust volatility rates, you are entering the area of forecast volatility. There are advisory services that publish volatility forecasts. And these numbers can be valuable guides, but they have the weakness associated with any analyst's attempting to foretell the future. Therefore, you should only use them as a guide, the same way you use historical volatility. Personally, I place more weight in historical numbers, particularly the highs, lows, and support and resistance levels. But always use the most updated charts. The ones in this book are meant for illustration purposes only.

MOVING AVERAGES

The analyst or trader seriously relying on volatility can use simple averages, weighted averages, or moving averages to get a better feel for what value to use when running simulations. Let's say you are planning to enter an option position. You want to run the Black-Scholes model, but you're not sure of what volatility factor to use.

Your first step should be to look at the historical values. The current historical volatility (over the last year or 250 days) is 15 percent, for example. Checking the last quarter or 60 days shows a figure of 22 percent, and the current figure, say the most recent 10 to 15 days, is 20 percent.

Volatility for the last quarter is 46 percent above the historical value. You should find out why. It could be a normal seasonal pattern, since the nearby rate appears to be returning to the historical level.

A simple average of the three values is 19 (15 + 22 + 20 = 57/3). This figure could be used. Or you could give more weight to the historical factor by calculating a weighted average. If you plan to buy and hold an option for the long term, you would give more weight to the historical perspective. Short-term traders would give the emphasis to the near term. For example:

Long Term:	Near Term:
15 × 5 = 75	15 × 1 = 15
22 × 3 = 66	22 × 3 = 66
20 × 1 = 20	20 × 5 = 100
161/9 = 17.8	181/9 = 20.1

Moving averages are an excellent tool for filtering out market noise and uncovering trends. Long-, medium-, and short-term moving averages can be compared. This often provides a valuable method for uncovering trend changes.

ESTIMATING VOLATILITY

So far, I have used historical forecasts and seasonal and implied volatility to get a fix on what volatility rate should be used in the Black-Scholes formula. Additionally, I have experimented with simple and weighted averages. Now I'll compare the volatility characteristics of related futures contracts.

Traders who specialize in trading certain commodity or stock sectors often chart the historical volatility of the entire group on a single graph. From a chart like this, you can more accurately gauge the relationship between the volatility rates of commodities or stocks in the group.

T-BILLS/BONDS

Don't forget that some futures contracts, like Treasury bills, Treasury bonds, and Eurodollars, for example, are indexed by 100. This must be taken into account when you calculate the daily or weekly volatility. To do this, you assume that interest rates will never be a negative number, just as you previously assumed that the price of a physical commodity would never go to zero or below. If T-bills were trading at 98 and the annual volatility was 16 percent, the daily volatility would be 0.98 (16 percent/16 × 98 = 0.98). This appears to be a bit unrealistic. Therefore, we use 100 – 98, or 2, in the equation (16 percent/16 × 2 = 2), which is more plausible.

SUMMARY

Volatility is the primary measure of how wide the price swings a futures contract or stock will make over a given period of time. An annual volatility rating of 20 indicates that the price of the entity being studied has the potential of ranging from plus 20 percent to minus 20 percent of the mean price, which is the current price for futures. For stocks, it is the current price plus dividends and carrying charges.

To make an intelligent decision regarding whether an option is oversold or overbought, and therefore whether it is a good investment opportunity, most option traders calculate a theoretical price. This price is compared with the current market price and trend.

To calculate the theoretical price of an option, you need to use the following inputs:

1. Time to expiration
2. Underlying futures or stock price
3. Strike price

4. Interest rate (dividend for stocks if paid before expiration)

5. Volatility

All of these are given except for volatility. Since volatility is the only factor that must be estimated, it becomes extremely important. Traders use historical records, comparisons, and other techniques to estimate the future rate of volatility. But no matter what approach is used, it is still an approximation. This again underscores the importance of understanding volatility.

One last note to emphasize just how important volatility is to the professional trader is the fact the Chicago Board of Options Exchange has developed a CBOE Volatility Index. This index measures the potential price variations in blue-chip securities traded on the U.S. stock markets and allows you to hedge the value of options on benchmark stocks caused by fluctuations in volatility. The S&P (Standard & Poor's) 100 Stock Index was used as the base because it is the most liquid option index in the world. A mathematical formula calculates the probability of prices moving 20 percentage points above or below a target price over a year's time. If you plan to trade stock options, you'll want to track this index. You can use it as a guide when you need to estimate volatility or as a checkpoint when you want to second-guess an estimation.

CHAPTER 6

Developing Trading and Money Management Plans

"When schemes are laid in advance, it is surprising how often the circumstances fit in with them."

Sir William Osler (1849–1919)

Key Concepts
- Putting Your Plan on Paper—Goal Setting
- Gut and Financial Checks
- Selecting Which Markets to Trade
- Trading Plan Checklist
- 9 Money Management Rules
- 12 Psychological Trading Rules
- 13 Trading Rules

Now you can get down to using what you've learned. So far, you should have a general overview of options trading, terminology, strategies, and approaches to price forecasting.

Where do you go from here? How do you get into the market?

First of all, I strongly recommend you develop a plan—and put that plan on paper! A written plan does two important things for you. It forces you to make concrete decisions, which I'll go over shortly. A written document also provides you with something you can measure your progress—and more importantly, your discipline—against.

It's sad the number of traders I've seen over the years who have waffled from one approach to another without any clearly stated plan. They wander in and out of the market, often making the same mistakes over and over.

Your written plan need not be fancy. It's not a formal proposal. It could be as simple as answers to the questions I'll shortly be discussing. But it will help you decide what you are going to do, how you are going to do it, what you need from outside sources, and what you expect to gain.

When you get to the point of selecting specific trades, try using the technique known as visualization. It first became popular in sports, particularly on the Olympic level. I use it to improve my golf game. You visualize a shot before you swing. In your mind's eye, you see the ball arc high, eventually dropping onto the green near the pin. With trading, you create the perfect trade in your mind based on your analysis. The yen moves higher for the next 30 days. Corn follows its seasonal pattern from its harvest lows. The rationale is that if you don't have a clear picture in your mind of where a market is going on the price charts, you're not ready to enter the trade.

Just like golf, not every shot or trade is perfect or near perfect. You won't shoot par every time any more than you'll make money on every trade. But in both cases, after you visualize how things should be, you know early on when they are not.

This technique helps you cut your losses short and lets your winners run for all they are worth. It will also be suggested that you keep a trading journal, which will be described in the next chapter.

WRITING YOUR PLAN

Step 1 is describing your trading philosophy and general goals. This can be done in one or two sentences. Here are a few examples.

Example 1
"I plan to trade for medium to long term, holding positions for 60 to 90 days. My primary strategy will be spreading."

Example 2
"I'm a short-term trader. I seek options in-the-money or near-in-the-money close to expiration and plan to hold these positions no longer than a few weeks."

Example 3

"My approach will be to write covered options. I'll concentrate on options one or two strike prices out-of-the-money."

Example 4

"I plan to hedge my stock portfolio. I'll buy puts in the S&P 500 when I think the market is in a bearish trend."

These are only examples, and reasonably conservative examples at that. A very aggressive trader could have written: "I plan to trade my brains out whatever the risk, trying to double or triple my money every month."

By jotting down your general approach to the market, you set the tone for the rest of the plan. It helps you answer the questions to come.

WHAT ARE YOUR GOALS?

We all want to make the most we can. If we didn't, we'd be looking at certificates of deposits or Treasury bonds. Your goals dictate your strategies and must reflect your philosophy. A goal of 200 or 400 percent return to equity per annum requires a different strategy than a goal of 25 to 50 percent per year. If these return objectives seem high to you, compared with other types of investments, remember options trading is very speculative. High return, high risk. No pain, no gain.

Let's go back to the point of matching investment goals with strategies. If your return objective is on the high side, say 200 percent per year, it is unrealistic or impossible to reach it using a "conservative" spread strategy, such as the buy-call spread. This type of spread, as illustrated in the strategy section, has a predefined maximum profit and predefined maximum loss. The maximum profit is no where near 200 percent. To have a shot at a 200 percent return requires a much more aggressive strategy, like buying in-the-money puts and calls. The risk, of course, is if the underlying entity trades sideways or in the wrong direction, your options will expire worthless.

Once your overall goals are set, you should break them down to per-trade goals. For example, let's say you are starting with $10,000 in trading equity. Your goal is to double it this year, or to earn a 100 percent return. That means you need to generate $10,000 in net profits over the next 12 months, or $834 per month.

If one of your money management rules is to never invest more than 5 percent of equity in any one position, you have at least twenty trades, or one or two per month. Let's use two trades per month, assuming that some will make money giving you the extra four trades per year. You, therefore, have two trades to make $834, or $417 each, net of commissions and fees. If each option you buy or sell takes $500 of your equity or the 5 percent, your basic approach is a dollar-per-dollar risk-to-reward ratio, which is low for options, or "conservative."

Additionally, you know in advance that not every trade will make money or even break even; some will be losers. For this reason, you look for trades with a 1:3 risk-to-reward ratio. The key to success is the discipline to focus on specific, achievable goals.

This line of thought brings us to the subject of risk. When attempting to measure or quantify risk, economists use the term "utility," which is a measure of personal satisfaction. If something generates a feeling of greater satisfaction than whatever it is compared with, it is said to have greater utility.

You need to seriously think about how much risk you are psychologically prepared to accept. Do you classify yourself as a person averse to risks, one neutral toward them, or an aggressive risk taker? For example, a risk-averse person may not be interested in a trade that projects a dollar return for every dollar at risk. These traders want to take little or no risk and are comfortable with low, but dependable, rewards. The buy-call spread and covered writing strategies appeal to them. The neutral and aggressive risk takers are willing to take greater and greater risks to get higher and higher utility.

Playing into this concept is your personal financial situation and responsibility. A single person with a handsome six-figure income can afford to take some risks. On the other hand, the breadwinner of a large family in a middle-income tax bracket should be more conservative.

As you probably guessed, you must factor into the analysis your personality. Will you be a good trader from a psychological standpoint? Options trading requires optimism, good judgment, and the ability to make good decisions under a lot of financial pressure. Excitable types rarely do well. Nor do worriers. Additionally, it helps to have a life free, as much as possible, from other major personal problems that can distract you from the markets.

After sorting through all these thoughts, it's time to set specific return goals that you will shoot for:

5 percent per week

10 percent per month

50 percent per quarter

150 percent per year

Set the amount and term to match your strategy. Long-term traders evaluate themselves on the long term. Short-term investors need to do their evaluation more often.

At this point, you are setting targets. You can revise these marks higher or lower the longer you trade. Within a year, you should have a really good feel for what you and your trading system can do.

WHICH MARKETS?

One of the next questions is markets, which ones? This decision starts with your price forecasting approach. If you anticipate using fundamental analysis, you should choose markets that you are very familiar with—ones that you understand intellectually and intuitively. This is because there are so many variables that can influence prices, there is no way to measure all of them. Even the most sophisticated econometric models come up short. Investors with farm backgrounds can trade the agricultural options, for example. Or serious students of the stock market should trade the stock index options.

If you're a technical analyst, you seek markets in which you can get signals that you are confident are reliable. Not every tech system works on every market or stock. Nor are technical systems dependable in every type of market, such as trending and nontrending.

The traders interested in using price volatility to spot overvalued or undervalued options, either as their sole analytic tool or as a supplement to another approach, should probably consider a computer program. As you'll see in Chapter 7 on automated trading (and in Appendix 4), several very good programs are available.

Related to the question of "which markets" is the question of how many different stocks or commodity markets you should trade at any one time. This is the concept of diversification. As you read the following, think back to the earlier discussion of the law of probability.

One of the benefits to selecting a full portfolio is diversifying your risk over as wide a spectrum as possible. The theory is both offensive and defensive. It's offensive in that it is designed to put you in the right place at the right time. By taking a position in several commodity markets or stock issues, you have a greater possibility of catching a major market move. No one—absolutely no one—can anticipate exactly when a given share or market will make its next big move up or down. Therefore, the more alternatives you are covering, the greater the chance you will be successful.

From a defensive stance, by spreading your investment capital over a wide variety of markets, you won't be caught in the wrong market at the wrong time—or even the right market at the wrong time. Option trading is a lot like fly fishing. You're constantly casting your line with the objective of making a big strike. If a trout doesn't hit, you reel your line in and try again. Option traders let their profits run and cut losses short. A good, successful trading system often has more losing trades than winners. The difference is that the overall gain from the winners exceeds the losses from the losers.

Additionally, experience has taught that different stock or commodity groups are normally negatively correlated. That means they do not move up and down in unison. When one group is bullish, another is bearish and a third may be moving sideways. For example, if grain prices are high (bullish) and going higher, that puts pressure on livestock producers because it cuts their profit margin. More livestock are subsequently sent to market sooner and livestock prices become bearish, moving lower as the cash market is flooded with animals. Understanding negative correlation can help you from getting caught with all your eggs in one basket.

Once you begin to reach some preliminary decisions on how, which, and how many options you can afford to trade, it's time to consider what you want to or can invest in options trading. In this section, we're talking in general terms. The rule of thumb is that the amount should not exceed 10 percent of your risk capital, which in turn is defined as money you could afford to lose without incurring a change in your lifestyle.

Experienced investors invest a small portion of their risk capital in highly speculative investments, like options, with the idea of generating a very high return. The purpose is twofold. First, the bulk of their net worth is held in more conservative investments for protection. And, second, the hope for a high return from options trading gives their overall return on investment a boost.

Here's a checklist of what questions should be answered in the introductory or overview section of your written trading plan.

1. What is my option trading philosophy? Am I a long-, medium-, or short-term trader?
2. What are my goals? Have I broken them down into per-month and per-trade goals? Are they realistic?
3. Am I psychologically and financially suited for options trading?
4. How do I plan on keeping a trading diary?
5. How do I plan to select trades? Close out or offset trades? What is my analysis technique?
6. What strategies suit my resources?
7. How many and which markets will I trade?
8. Are the answers to all of the above questions consistent?

As you write your plan, keep in mind that it is a living document. It will change as your experience increases, as markets change, and as new option products are introduced.

You should also begin at this stage to prepare a list of your limitations and needs. This can be anything from something personal, like discipline, to something physical, like quotation equipment, to something intellectual, like research on the underlying entities. Later, I'll show you how to use this list of shortcomings in your selection of a broker and brokerage firm.

Once you've written general, almost philosophical, answers to the eight questions listed above, you should prepare a list of specific trading rules to live by and trade by. The purpose of these rules is to instill discipline into your trading. Your trading decisions are business decisions—treat them with the same cold logic you would use to address the refinancing of your home or business.

To help organize these rules, I have divided them into three general categories—money, mind, and management. You need to sort through my suggestions and extract the ones that particularly pertain to you. Then tailor them to your financial, personal, and trading style.

Of the three groups, I personally believe the money management principles are the most important. My reasoning reflects the law of probability already discussed. Good money management allows you to survive the losing trades you will invariably encounter so you will still be trading when you are in the right market, with the right strategy at the right time.

9 MONEY MANAGEMENT RULES

1. The first and simplest thing to do is divide your risk capital into at least 10 parts, preferably 20. To put the law of probability at your back, you want to be able to make 10 or more trades. You're attempting to have the normal distribution of winning and losing trades make you a net winner dollarwise. If you do not have enough risk capital to make 10 trades with a reasonably good chance of succeeding, seriously consider waiting until you do. If, for example, you only have $1,000 in risk capital and you decide to buy 10 deep-out-of-the-money options, the odds are so far out of your favor you would probably be better off at Churchill Downs. Buying 10 deep-out-of-the-money options is not "investing" in options, in my opinion.

2. Always calculate the transaction costs as part of the risk-to-reward ratio. The major component of this is the brokerage commission. Don't overlook it. If you overtrade your account, commissions can draw down your equity, just like losing trades do. I'll discuss the negotiation of brokerage commissions in detail later.

3. Never forget what you are trading. Options are wasting assets—eventually they will expire worthless unless offset or exercised. Think of your options as 100-pound blocks of ice. The time to expiration is the ambient air temperature. The food in the icebox is the portfolio of underlying entities. The purpose of the ice is to keep the stock or commodity options fresh, until you are ready to use them. But the longer you wait, the more the ice melts and the less time they'll stay fresh. The closer you get to expiration, the faster the ice melts, as the block of ice gets smaller. At expiration if you don't do anything with the options, you'll have a puddle of warm water and a pile of "spoiled" underlying entities. In other words, you forfeit 100 percent of the premium you paid for the options.

To conserve your risk capital, offset or exercise before expiration! All too many new traders allow options to expire worthless—pros rarely do.

4. Always know what your risk-to-reward ratio is. Divide what you expect to make in profits from an option by your total cost (premium, commissions, fees). If it is not at least 3:1, you may want to look for a better opportunity. Exceptions can be certain strategies, like the buy-sell spreads, that have a fixed downside risk with high probability of reaching the upside objective.

5. Be very careful of pyramiding. This is the practice of buying additional positions with unrealized profits. It increases your leveraging, but also increases your risk.

6. Avoid letting a profitable position become a loser. This sounds obvious, but it is a common mistake of new traders. It is often caused by a trader taking his or her own analysis too literally.

Here's what can happen. A trader forecasts a move in an underlying entity (stock or commodity) and buys an in-the-money option. The move occurs as predicted, but ahead of schedule or with more momentum than expected. Instead of placing a trailing stop-loss order to protect profits or promptly offsetting, the trader holds the position too long waiting for a certain target price to be reached. While waiting, the positions make a negative move and the option eventually expires worthless. Markets usually fall twice as fast as they rise, as a rule of thumb.

7. Diversification of positions into several different classifications should be included in your money management rules. For example, you might have a rule that no more than 25 percent of your positions are in any one stock or commodity group.

8. Don't forget the biblical parable about seven lean years following the seven bountiful years. Options trading often follows a similar pattern. You find yourself with a string of winners, only to be followed by a string of losers. My recommendation is to siphon off profits to an emergency account, when available.

Another good approach is to withdraw your original investment from your trading account as soon as possible. Then you are trading with profits only. This sometimes relieves much of the pressures traders feel when they lose their "own" money.

If you can withdraw some profits, on top of your original investment as well, it helps you stick to your original plan, which obviously was successful. Do not automatically reinvest all profits in the market. A "go for broke" mentality is very dangerous in this arena.

When a losing streak begins, take a trading vacation or cut the size of your trades in half. If you are trading 10 lots of a certain spread, reduce trades to five lots. If losing trades continue, cut down to one or two lots. Eventually, you'll recover, but you can't if you have traded away all your equity.

9. From time to time, you may be approached with a strategy that makes good sense from a tax situation. Do not let this be the reason for taking the trade. Take the time to do your regular analysis to assure yourself it is a good trade on its own merits.

12 PSYCHOLOGICAL TRADING RULES

Now let's discuss some psychological rules you might want to incorporate into your written trading plan. To paraphrase a famous baseball saying: "Options trading is 90 percent mental, and the other 50 percent is knowledge, experience, and skill."

1. Rule numero uno is to follow all the money management rules. Money at risk is, of course, a very serious source of stress. If those rules are violated, following the ones described below will not do you much good.

2. Experiencing losses in the option markets is not failure. It's the normal course of business. You can equate it to trying to hit a baseball. A hitter who has a .300 batting average gets a hit about every third time at bat. Two-thirds of the time he loses. Worse yet, of the three in ten hits, only a small percentage of them are home runs. An even smaller percentage of the home runs are grand slammers. A lot of the hits are singles.

In option terms, singles break even and may pay, or partially pay, the transaction costs (commissions and fees). The batter who hits .300 or has a lifetime average that high will be a success. So will the options trader whose every third trade is profitable. The law of probability will see to it that he or she gets a few homers and an occasional grand slam.

3. Don't run your mind through the wringer on every trade. Once you make up your mind on a position, stay with it, unless you uncover some overwhelming reasons to abort the trade. Avoid listening to the advice of others once your mind is made up.

4. Closely related to rule 3 is the ability to live with your trading decisions. You must accept full responsibility for them and that's that!

5. At the same time you're accepting responsibility to live with your trading decisions, stay impersonal. Be prepared to cut losing positions early through the use of stops.

6. When you feel the market is dragging you down, walk away for awhile. Take a vacation from the markets.

7. Take pride in yourself and your trading skills. It is the only thing that keeps you going sometimes—the market can be a "great humbler." Therefore, you need to control your reaction to success and stay humble. Pride often induces overtrading, which can result in serious financial damage.

8. Trade as James Bond would—with cold, calculated precision, once you have decided what to do.

9. Maintain the maximum amount of self-discipline. Set concrete rules for yourself—for example, stop trading for a day or a week after taking two consecutive losses—and stick to them.

10. Master yourself. As part of your diary, include remarks about your behavior. Then work at correcting any flaws. You must work against natural instincts to be stubborn, emotional, and inflexible. Find your character flaws and overcome them.

11. Be well aware of who you are. Know your level of risk tolerance; you won't make sound decisions when you're way over your head financially. Nor when you're too tired or ill or not prepared to trade. Don't be afraid of missing an opportunity by taking a day off. Believe it or not, there are always great new trades in the offing. Learn to be patient; pick and choose your trades carefully.

Match your personality and temperament to your trading system. If you're not a detailed person, use a system that requires as little detail as reasonable. If you're very excitable, avoid day trading. Consider being a position trader, and once you put a position on, let it alone until you're ready to offset it.

12. Try not to overtax your resources, particularly time. Carefully evaluate how much of a commitment in time, money, and mental resources you can make. Part-timers should limit themselves to tracking three to five markets or stocks and trading only two to three. The markets should be ones that you have some prior experience with, have a special interest in, or are knowledgeable of through your profession or life experience. For example, traders with farm backgrounds often have a good feel for the agricultural markets. Trading markets you're fond of makes the research easier.

13 TRADING RULES

The last set of rules deals with actual trading. Here are some suggestions of certain things you should or should not do. These axioms come from the real world of options trading.

1. This rule contradicts all the ones that follow—be a Doubting Thomas!

First off, don't believe anything you hear about trading. Use all you learn as a guide, and find out what works for you as an individual. This frame of mind spills over into what your broker and fellow traders tell you is going to happen in a particular market, or their advice in general.

I have seen a lot of people make a lot of money in the options markets, but I've seen only a few who made money consistently by listening to someone else. You need to distinguish between useful information about the markets and someone's opinion. The basic rules, combined with your education, experience, and understanding of your markets, blend into a unique approach. This is your trading system, which includes money management and psychological restraints, as well as your trade entry and exit decisions.

2. Be totally open—to new ideas, new strategies, new markets, new information, new forms of trading and communicating. Never forget you're attempting to predict what will occur at some point in the future. Most traders assume price patterns repeat, but exactly when and how closely the patterns mirror past ones is unknown. Look for the old, but trade the new.

Your approach to the markets should include the flexibility to change your mind, to trade the short side of a market as well as the long, to make adjustments in your money management or trading system, and to consider markets you haven't previously traded. A successful trader is often the one aware of the most alternative paths. He or she has the flexibility of a master chess player to adjust to the unexpected.

3. Place your markets in their historical perspectives. Even if you don't use them as signals or indicators, you should know the historical highs and lows of the markets you trade. Additionally, you must have a good understanding of the seasonal patterns and major cycles. Why? Simply because a large number of technical traders will trade based on these patterns. You must take their anticipated reaction into account as part of your overall trading analysis. If, for example, you buy a corn put at what are the normal seasonal lows (at harvest), you better have good reasons or deep pockets.

4. The same reasoning expostulated above applies to developing a sound knowledge of technical analysis. It is often called a "self-fulfilling prophecy" because so many traders see the same technical signal and react in unison. For example, an uptrend line is broken, alerting technical traders that the current uptrend is now a downtrend. Just like a herd of longhorns heading for water, they stampede to the short side of the market.

5. Don't spit into the wind, trust a junkyard dog, or trade contrary to the trend. There are good financial and psychological reasons trends are created and maintained. For instance, would you pay more for something you intended to resell if you didn't think it was going to increase in value? Of course not. You buy a call because you believe you'll be able to sell it for more in the future. The same is true of a downtrending market. You buy a put because you think you'll be able to offset your positions at a lower price.

The amazing thing is that this is nothing but human nature. We all feel this way. Just about all traders trade this way, causing markets to trend.

Therefore, you need to know the trend of the market you are about to enter before you open a position. You can go either with or against the trend. If you go against the trend, you need a good reason—such as that it is about to enter an area of resistance you feel it cannot penetrate, or you become aware of some fundamental information that is not yet in the market but has the potential of turning it around. Either way, you want to know what the herd is thinking. This is the trend.

6. Develop a definite plan or strategy for entering and exiting trades. For example, you might enter a market with limit orders and exit by continually moving your trailing stops closer to the trading range until they are hit. The danger you want to avoid is not carefully planning entry and exit points. All too often, traders get in markets too late and exit too early. Or novice traders get filled "unexpectedly" and have no idea what to do next, nor have they any strategy planned to exit the trade. They just let their options expire worthless.

7. A corollary to the previous rule is avoid closing positions prematurely. I've seen experienced traders very orderly open a position and precisely close it when their price objective was hit. I used to tell my customers: "You can't tell how far a market is going up (or down) by how far it has gone already." The only true way of knowing is when it changes direction and is no longer going up or down. Following positions with protective stops also corrects this problem.

8. Learn the art of placing protective stops. It is easy to tell you to place stops, but it is not as easy to tell you how or where. It requires much experience with the specific markets you trade and a good feel for the current and expected volatility. You need to place your stops close enough to where your position is trading to protect yourself from losing what you've gained so far, yet not so close that you get prematurely stopped out. A good place is just beyond a technical barrier and outside the daily maximum volatility target. Avoid points known to be the target of floor traders because these traders will stretch the trading range to reach the orders in their deck, especially in thinly traded markets.

Stops are particularly important in options trading because you are trading a wasting asset, which will expire worthless if you don't do something. Let the stop salvage something out of a losing position, if the markets move against your positions. Don't forget what I warned you about earlier regarding stops in illiquid markets. Use mental stops or give your broker some discretion if there is low volume in the futures market or the stock you are trading.

9. Plan and develop rules for taking profits. For example, don't exit a market just because you've become bored trading it. On the other hand, if a market moves dramatically in your direction and you have no idea why, take your windfall profit or at least put a tight stop behind the position. You want to avoid any situation you don't understand. You'll never forgive yourself if you take a serious drawdown on a trade you didn't understand.

10. Learn to trade puts as well as calls. Most traders buy only calls. They neglect the short side of the market. The high number of trades on the long side of the market is particularly noticeable among stock option traders. The reason for this may be that they feel guilty about "letting down a company" if they trade the downside. They forget to realize that the company is not directly involved. It's the owners of the shares of stock that are affected. The company only gets money from stock sales during the initial offering or when it issues more stock.

11. Never "hedge-trade" or try to spread out of losses. True hedging occurs when you are on both sides of a market. The correct way to do this is to be on one side of a cash market and on the other side in options or futures. For example, a farmer has 5,000 bushels of soybeans in a storage bin or growing in her fields. In other words, she is long 5,000 bushels. To hedge, she sells 5,000 bushels or buys a put option on the Chicago Board of Trade. She is now neutral pricewise. When the commodity contract is about to expire, she can deliver the beans or offset her position. The rationale for hedging is to pass the risk of ownership to a speculator and lock in an acceptable profit.

Some option traders, when faced with a losing long position, buy a put at the same strike price on another exchange. They call it hedging, but it isn't. It's more like an intermarket spread. This is bad trading. When faced with a losing position, offset it and take your loss. If you wish to trade spreads, do so. Look for spread opportunities when the relationship between two related contracts is, by your analysis, misaligned. If you decide to reverse your position, do it. But do it as a completely new trade.

12. Never buy a call just because the underlying entity's price is low, or a put because a price seems high. Traders often get trapped by the logic that if the underlying entity is very low, it must go higher. This may be true in the long run, but it sure isn't in the short to medium term.

Remember the old onion futures contract that went below the cost of the bags in which the onions were packaged. Also, silver is often a by-product of lead and other base metal mining. Its production can continue after its price drops below production costs. Even ranchers continue to produce livestock when in the red because it takes 2½ years from the time a cow is bred until a steak is produced from its offspring. This process can't be turned on and off at will.

Another example relates to the silver market in early 1980. Suppose you shorted (bought a put) the silver market in early 1980 because silver was "too high" at $30 an ounce. You would have cried when this commodity shot up another $10 more an ounce. That's $50,000 more profit per contract.

13. Never leg out of a spread, unless this is your original strategy. Legging out of a spread means selling off one side and continuing to hold the other. You should be in a legitimate spread for legitimate reasons. Therefore, when you close out this position, you should off-set both legs.

I will summarize this section by emphasizing the biggest "secret" of options trading—it is self-discipline, the ultimate key to successful options trading. If you learn nothing else from this material, this is the most important single aspect. Here are a few tips:

- Create a comprehensive trading strategy. This includes money management as well as a trading system. Put it in writing.
- Money management is the more important of the two.
- Never trade when you are sick—physically or psychologically.
- Follow your own trading system. Avoid the advice of others. If you want someone else to make your trading decisions, give them the discretion (limited power of attorney) to do it.
- Only trade your own trades. If you are trading someone else's, you'll lack the conviction to stick with them if the sledding gets tough. And only make trades that are clearly defined, both the entry and exit points.
- Set break periods. For example, if you have three or five losses in a row, it's time to back off for a while.
- Keep a journal explaining why you entered each trade, why you exited, the results, and a critique of your performance. Compare what you actually do to your initial money management and trading strategy. I'll have more advice regarding your journal in the next chapter.

There is one more way to think about options trading. It is summed up in an old market axiom: "The markets are always right!" In other words, they move up, down, or sideways *despite* what any analysis predicts. You need to be disciplined enough to take advantage of what the market gives you. I'll have a little more to say about discipline shortly, because it is such a key to successful options trading.

CHAPTER 7

What You Need to Do to Implement Your Trading Plan

"This very remarkable man
Commends a most practical plan:
You can do what you want
If you don't think you can't,
So don't think you can't think you can."

Charles Inge
(Weekend Book, *1928*)

Key Concepts

- What Is a Trading System?
- Types of Systems
- Discipline and Your Trading System
- The Value and Use of a Trading Journal
- Dealing with Your Pride, Greed, Fear, and Hope

After reading through the lists of rules, the chapter on price forecasting, and the section on strategy, you might be wondering what a trading system actually is? How do you put one together, and what do you need to do to make it successful?

These are not simple questions, and there are no definitive answers. To properly deal with them, I need to digress for a minute.

131

In the chapter on volatility, I discussed the random nature of prices and several times I made reference to the law of probability. These two concepts converge when you attempt to develop a trading system.

First, you must resolve in your mind the dichotomy between random price movement and your ability to forecast prices into the future. If price changes are truly random, what good is a trading system based on the expectation of knowing at least the direction in which prices will go, not to mention being able to predict specific target prices days or months into the future? And all systems do attempt this in some degree or fashion.

Here's how to deal with this dilemma. I hope I made it clear earlier that neither technical nor fundamental analysis is anywhere near 100 percent accurate. Additionally, each specific analytic technique is further limited based on the specific characteristics of each underlying entity and the trading pattern of the market being studied. For example, a character-of-market technical forecasting method does better in choppy markets than a trend-following system.

My approach is to pick and choose the technique based on the underlying entity and market conditions. It is not uncommon to use more than one, even several, at the same time. You get a signal from a price chart and look to an RSI (relative strength index) for confirmation. Or you may use fundamental or chart analysis to isolate a trend and Japanese candlesticks for market entry and exit points.

You need to think of your analysis as an estimate. If price changes are truly random, it is an estimate or an educated guess. If price changes are not random—which implies they follow discoverable patterns or rules—why haven't extremely reliable trading systems been developed? Carrying this logic farther, if dependable forecasting approaches can be uncovered, there would be no need for option markets, nor stock or futures markets. More importantly, if these laws governing price activity can be uncovered, why not the laws of human behavior? No need to stop here, why not the future? People have been attempting to foretell it since verbal communications began—but to no avail. Despite what ardent promoters of seasonal, cyclical, Elliott wave, Gann numbers, and hundreds of other analytical approaches say, their forecasts are only educated guesses at best.

If you develop a way to accurately forecast the future, would you share it? Would you sell it to all comers for $99.95, $495, $10,000, or $100,000? If everyone had the system, would it still work? Would you sell it and take that risk?

In other words, the ability to foretell the future is not currently available; at least I can't share it with you at the moment. This is where your price forecasting approach becomes part of your total trading system.

Since your opinion of the direction of prices is only an estimate, you need to learn how to manage the risk of investing in your market opinion. This is done in many ways. The use of trading techniques, such as stop-loss orders, is one. Refer to the money and trade rules for others.

Another important part of this defensive stance is selection of strategy. As you saw in the discussion of specific types of trades, some are more conservative than others. Or you could say some are more risky.

At the point where you have settled on a specific trade, you need one more input from your trading system. I am referring to entry and exit points. How (and when) do you plan to get into and out of your trades? This must be addressed before trades are put on. You can always make adjustments as conditions change, particularly if they are favorable to your position. I'm just saying that the most successful traders I know trade with a definite plan in mind—better yet, on paper.

Try thinking of your trading system as a three-legged table. One of the legs is your opinion of market direction, or if an option is overpriced or underpriced; leg 2 is management, monetary and psychological; and leg 3 is the mechanics of the system, meaning the selection of such things as entry and exit points. To be structurally sound, all legs must contribute to the support of the table. If one of the legs comes up short, the system topples and falls.

WHAT TYPE OF SYSTEM IS RIGHT FOR YOU?

There are basically two types of systems: mechanical and intuitive (often referred to as discretionary).

A 100 percent mechanical system would theoretically make all the decisions for you. You would incorporate all your rules for money management, stock/commodity market analyses, and trade selection. The system takes all the data you feed it and "tells" you the best opportunity.

A purely intuitive approach is just the opposite. Unfortunately, too many traders select trades this way. They just do what they feel like doing or what their brokers get them excited about. Occasionally this approach works for very experienced traders.

Most of the successful individual traders that I have dealt with find a middle ground between these extremes. They will, for example, have a set of money management rules they never violate. For trade selection, they may use a computer program or some other type of mechanical process. The mechanical system analyzes the current markets, selects the best opportunities, runs breakeven tables, runs risk array scenarios, and ranks the trades indicated based on risk-reward criteria. This information is combined with experience (which often translates to looking at dependable patterns in the fundamental or supply-demand situation, seasonals, or cycles), generating the trade decision.

To make this type of system, or any other, work requires a good deal of discipline. By discipline I mean training intended to produce a specific pattern of behavior or, more simply, controlled behavior. You want to train yourself to do something as nearly perfectly as possible.

Let's forget trading for a moment and examine the concept of discipline from another perspective. What if you want to lose 30, 40, or 50 pounds? Knock 6 inches off your waist? Go down two or three dress or suit sizes? More importantly, once you lose the weight, you want to keep it off.

What steps would you have to take? Cut down on your caloric intake and increase the number of calories you burn each day. Eat less and exercise more. That takes care of the first goal of losing 40 pounds.

To successfully maintain the new lower weight, you have to permanently change your eating patterns. To do this, you're talking about a major commitment. It's one thing to give up chocolate ice cream for 30, 60, or 90 days. It's quite another to give it up forever. That's behavior modification!

Success in this endeavor also requires you to become more specific and follow up on your progress. This calls for written goals and menus. Then you must weigh yourself regularly, record the results, and honestly evaluate how you're doing.

Most people gain back what they lose because of lack of commitment and discipline. Most option traders do the opposite—they give back what they gain for the same reason. Written goals, written plans, follow-up procedures—that's all fine, but without the wholehearted commitment to making money from the options markets, it means little.

Now, let's get back to the three areas in which you, as an option trader, must train your behavior. The first is your trading system. Studies seem to indicate there are a wide variety of systems that can make money in the markets.

All trading systems have a few functions in common. They usually give specific trade entry and exit signals, along with some type of a timing mechanism. Your function is to take the appropriate action.

If this is carried to the limit, the ideal trading system would be totally automatic. You'd hook it up to a computer order-placing system and have the profits wired to your bank.

However, I don't think this would work, because even the most automatic systems need the human touch to adjust to the idiosyncrasies of the market. A good analogy might be a plane with the guidance system of the cruise missile. It's great as long as the ride is smooth, but what happens when the hydraulic system blows or an engine stalls? You sure want a pilot or two on board then.

On the other hand, the human touch must be gentle and timely, or we face the dreaded "operator error" syndrome. It's discipline that keeps us on the straight and narrow. Discipline also makes sure you do your homework—updating the system regularly, reviewing it, and following the signals.

Money management is the second critical area. I hit this hard when I discussed the 9 money management rules. The only thing needed here is to be sure you adopt those rules that apply to your trading style. Put them in writing and keep them in front of you.

The last area refers to personal performance. You must discipline your emotions—particularly your pride, greed, hope, and fear, which I'll discuss shortly. These are the emotions that lead you back to your old losing ways.

HOW DO YOU STAY ON TRACK?

You begin with honesty. If you want to lose weight, you need to say to yourself: "I'm fat and I'm going to lose 40 pounds in the next 3 months." If you want to win in the futures market, you must say: "I'm going to enter some losing trades—there's no avoiding that. But I'm not going to let that bother me because I'm going to follow my trading system's rules, my money management rules, and my personal emotional rules. I'm totally committed to making money in the options market and keeping it!"

From here, you develop specific strategies to accomplish your mission. For example, it is easier to follow rules if you set a firm schedule. Update and review charts daily by 7:00 p.m. every trading day. Discuss markets with your broker at 8:00 p.m. each night. Think about recommendations overnight and call your broker in the morning to place orders. Update your trade journal before leaving for work. Check the markets at noon.

Back up your schedule with checklists, so nothing gets overlooked. It helps to match your homework with your personality. If you are not the type to do a lot of detailed number crunching each night, you need a simple system that doesn't require it or can be automated. If you flat out don't have any time, you shouldn't be trading.

Successful options trading requires a lot of work on somebody's part. If you aren't up to it, consider an alternative approach to the markets, like an account managed by a professional trader or a fund, if you feel you absolutely have to be in the options market. But remember, you don't!

FOLLOW YOUR ROAD MAP

Your trading system and written plan is your road map to success in the options market. Discipline is your navigator that keeps you on course. Be sure to listen to your navigator—so you'll end up where you want to be.

One of the best ways to assure you're on the interstate highway system to success is to keep a trading journal. Unfortunately far too many traders neglect this valuable tool. They often say: "Why should I bother? Don't I get a monthly statement recapping all my trading activity?"

The most important piece of information missing from your monthly statement is the reason you got into the trade in the first place. What did you expect to happen? What was the risk-to-reward ratio? What alternative trades did you turn down in favor of the ones you took? What signals—fundamental or technical—triggered your trading decision? What attracted you to the trade initially?

How does the old adage go? "Those that forget the past are condemned to repeat it." Trading is an extremely personal and emotional endeavor. Our minds interpret the ink blots on the price charts. We learn to rationalize our trading decisions. Most good traders talk about their instincts—they can smell a trade when it is still 10 ticks away!

The essence of what these traders are saying is basically believable. But we also know from our experience that writing stimulates thought and clarifies thinking. Researchers who have studied the papers of some of the world-class traders have indicated that their trading wasn't done by the "seat-of-the-pants." The great traders documented what they did—before, while, and after the markets were open.

We learn from our mistakes. Why repeat them? We learn from our successes. Why not repeat them?

The function of a journal is twofold. First, it reminds us why we selected or rejected a trade. It provides an audit trail of our thinking.

Second, we need this historical record because it helps us remember what really happened. We all have a tendency to mold the past by our perception of the present. If we try to re-create our thinking and our emotional state after the fact, we find ourselves heavily influenced by the results of the trade. This confuses us to a point where we can't be brutally honest.

For these two reasons, you should put your thoughts and emotional reasons for each trade on paper before you call your broker. This is needed even if all you have to go by is a "gut feeling that corn is going higher." Record when you first noticed that feeling and what might have triggered it—a news story, a chart formation, etc. Later on, you may be able to piece the puzzle together and make something more concrete of it.

Also, if your broker talks you out of a trade by supplying you with some information you didn't have, make a note of that. Follow the trade. Evaluate your broker's advice. Keep a mini track record on him or her. This will provide you with a concrete evaluation. Brokers, like the rest of us, have a better memory for success than for failure. If you decide to change brokers because the advice or information given is weak and the broker attempts to talk you out of moving your account, you have something in black and white to remind you exactly why you're changing brokers. Remember, the ability of a broker to earn a living depends on the ability of a broker to be persuasive.

Reliable instincts are as meaningful in options trading as they are in any human activity. The price trends and patterns of the stock and futures market reflect the instincts and emotional state of every investor who is getting in or out of the market at a given time. You may be able instinctively to sense where a market is going at times, just like fishermen can sense where the fish are.

Your trading journal helps you hone your instincts. You learn to sort out unfounded hunches from brilliant insights. Fine-tune your feel for the market through the use of a journal, just as the professional fishing guides chart their success on the lakes they fish.

WHAT–WHEN–WHERE?

There are three important questions. What should you include in your trading diary? When do you need to make entries? And where do you keep your diary?

The diary reflects your trading system. To determine what to include, think about your answers to these questions: How do you select trades? What are your key indicators? What are your goals? For example, the first thing many traders do is determine if the trend is up or down for the markets they are tracking. This may be your first entry. You would then do the same for the other signals that you regularly check.

Or you could have your signals listed and just make a checkmark or other sign. You could use an "up" arrow for an uptrend and a "down" arrow for a downtrend, for example.

How often do you need to make an entry or update your journal? There are a couple of ways to look at this. First, do you normally trade for the long or short term? Do you trade slow-moving markets, like the blue-chip stocks, or speed demons, like the S&P 500? You should make an entry each time you "check" your markets—in other words, whenever you evaluate your current position or seriously consider entering or exiting a trade.

What format is best for a trading diary? I've seen all kinds that work, ranging from hieroglyphic notes scratched on a Daytimer to an elaborate record maintained on a computerized word processing program.

The what, when, where, how, etc., of keeping a journal is a lot like the question of how long should a person's legs be? Long enough to reach the ground. Your journal should satisfy your needs. It is created only to help you trade more successfully. The worst thing it can become is a chore. If you find yourself dreading the thought of updating it, something is wrong. But don't, on the other hand, rationalize not doing one.

Simplicity often makes diaries more meaningful. For example, let's say you receive a weekly charting service, which you update once or twice throughout the week. You could make your journal entries right on the charts as you are filling in the daily trading ranges. Then cut out the charts you use and put them in a three-ring binder. Once a trade is closed, record the results. Then page back through the chart pages and evaluate your performance. On the last page, write a two- or three-sentence summary.

The ultimate goal of most traders is to make money. You want to be a net winner. Why not track your progress?

Start by dividing a sheet of graph paper into trading periods. It could be 12, 1 for each month, or 52 for the weeks of the year, or 250 to denote the average number of trading days a year. Your selection depends on your level of trading. What I'm suggesting is that you plot the "net liquidating value" of your account. You can do it each time you receive a transaction statement from your broker, or you can use your monthly statements. Footnotes at the bottom of the graph can be used to record additions and withdrawals. One hopes that there will be only one initial funding followed by steady growth and withdrawals.

A graphic representation of your equity often gives you a clearer insight into your progress. Additionally, the procedure sharpens your focus on your goal.

I also strongly suggest you treat each calendar year as a completely new year. Carry over your year-end equity and begin figuring your gains or losses from month 1. The reason for this is that if you had a net losing year, you now have a fresh start. Or if the past year's trading was profitable, particularly very profitable, you don't get overconfident because you have a big cushion of profits. Stay lean and mean.

The key is making your diary simple, painless, and meaningful. Your objective is to learn how you pick a winner and how you can avoid losers.

Some of the most difficult obstacles for new traders to deal with are pride, greed, fear, and hope. Some market observers, with literary bents, have likened them to the Fates of Greek mythology or the witches of Macbeth.

WHAT CAN OEDIPUS REX TEACH YOU?

The ancient Greeks would easily relate to how an intelligent, aggressive option trader—with a strong market position—ends up taking a massive, net loser. The Greeks coined a word for what often happens to otherwise skillful traders. It was "hubris," the common definition of which is simply "overbearing pride." You will talk to a trader about the risk of his or her position only to get a response like—"Don't worry about me. I KNOW where the market is going!"

You'll often find three witches accompanying hubris. They are hope, fear, and greed. I personally believe this foursome has brought more people, who could have been successful in the market, to their financial knees than just about anything else.

HOW DO YOU CONTROL THEM?

The answer may also be hidden in the past. Just as the Ancients offered an ox or a goat to the gods, you must make sacrifices. We often refer to it as "swallowing one's pride." One way of doing this is being brutally honest with yourself in your diary.

Let's say your system flashes a sell signal for a long position when an area of resistance is reached. You override it, only to watch your profits evaporate as prices head to the next lower area of support. In your diary—if it is to be of any real value to you—you must state how you violated your system. Your journal must force discipline upon your trading. If it does only this, it will be well worth all the time and trouble it takes to maintain.

Sacrifices must be made to the witches as well. Let's take greed as an example. You must prove to the market that you have "conquered" or at least can manage greed.

How can you do this? You do it by never letting a profitable position become a losing position. Let's say you bought an in-the-money silver call. Your expectations were to make 10-cents-per-ounce profit before expiration, which was 6 weeks away when you placed the trade. Silver unexpectedly moved 25 cents higher almost immediately after you are filled. What do you do?

(a) Take your unexpected profit?

(b) Wait for it to go up farther?

(c) Place a close stop to protect profits?

If you selected (b), you're testing the gods. And the gods and the markets punish mere mortals who challenge their authority. If you chose (a), you pocket $1,250 (less transaction costs) and the gods are pleased you accepted their gift.

Those of you who selected (c) are still fighting greed. But you have shown at least some management. You're still in the market. You still have a profit, at least on paper. If the gods smile on you, you'll eventually be stopped out with your profit. If not, the market will limit through your stop and plunge lower until you see the evil of your ways.

If you trade multiple positions, you can demonstrate your management of greed to the gods by closing out part of your position when it becomes profitable. For example, let's say to trade in three lots. You are holding three calls. The market moves up to one-half to two-thirds of your "final" objective. You close out one position at this point. Then you close your second when your objective is hit, and you speculate with the third using a trailing stop. This approach should please you and the deities that watch over these matters.

"HOPE SPRINGS DISASTER!"

Eternal hope can put you in the poorhouse. You may survive on hope in your personal life, but it can be extremely damaging in the markets. The usual scenario evolves around a position that is moving against all your best judgments. Just about everything indicates you are wrong—but you keep hoping the market will turn in your favor.

Hope's witches brew is addictive, numbing your discipline. Open your diary and write down why you are staying in this market. If you're frank with yourself and write "because I hope it will go my way tomorrow," you're in trouble. Cut your losses short. If you still think the market will turn, place a stop reversal order in the market. But do it only after reevaluating the situation and entering it as a completely new trade that meets all your criteria for a new trade.

The last witch is fear. It's one we face daily in the markets. Fear of losing money. Fear of making a stupid error. Fear of looking foolish in the eyes of our peers or broker. Fear, fear, fear.

Probably the most effective way of controlling fear is by sacrificing a portion of your profit. The adage—"Bears win, bulls win, hogs lose"—was born of those that have learned to deal with fear. For example, you can limit some market fear by not trying to pick tops or bottoms. If you are riding a bull market to seventh heaven, get off at the sixth cloud. Take your profits and bank them.

Another suggestion is to continually take excess profits out of your account. Then, when you do have a drawdown, it is not as crushing. You still have profits to show. You don't feel as threatened, and fear doesn't overwhelm you. Plus when you must write a check to put money back in the market, it reminds you of the fact that it is a two-way street. This is something we all need to be reminded of from time to time.

New traders often disregard fear, until they have their first serious setback. This can occur even when they make money overall on a trade. For example, a trade makes a spectacular move, generating a profit of $10,000. Before it can be offset, the market reverses and the trader gives $5,000 back to the market. It happens so fast the trader has problems adjusting. For the first time, this person realizes money can be lost just as fast as or faster than it can be made. "The market giveth and the market taketh away." You must be psychologically prepared to deal with the fear of being financially burned at some point in your trading career.

CHAPTER 8

What a Difference the Computer Chip Has Made

"To err is human but to really foul things up requires a computer."

Farmers' Almanac *for 1978*

Key Concepts

- ◆ Reasons for Using a Computer
- ◆ Danger of Relying on Computers
- ◆ Basic Types of Trading Programs
- ◆ How to Buy Software
- ◆ The Impact of the Chip on Market Trends

Trading Software

There are some very compelling reasons to consider making the computer a part of your trading system. Here are just a few:

1. *Discipline.* A good computerized trading system helps you tame your impulses. Successful traders, in my experience, rely on their system and follow its rules. They are controlled and systematic, not wild and unrestrained. Software is no substitute for a stable personality, but at times it can slow down someone who is overly emotional.

2. *Education.* Many of the software packages do an excellent job of teaching traders how the market reacted to past events, providing insight into what could happen in the future.

3. *System testing.* If you have an idea how you'd like to trade a market, you can simulate past markets and evaluate or fine-tune your approach.

4. *Speed/saving time.* Properly managed and structured, a computerized system can reduce, for example, the time it takes to update 40 markets each day from hours to minutes. Another good example is the calculation and recalculation of complex equations, like the Black-Scholes formula for theoretical option prices.

5. *Historical perspective.* With a few keystrokes on many systems, you can view a long-term or short-term chart of a specific market. This can often be done simultaneously with multiple windows on a single screen.

6. *Ease of analysis.* Sometimes it is just easier to do certain types of analysis on a computer. Examples would be drawing Gann lines and squares and calling up the meaning of Japanese candlesticks formations.

7. *Second guessing.* Some traders review various computerized studies to make sure they aren't missing an important signal.

8. *Paranoia.* Technical analysis can be a self-fulfilling prophecy. A quick review of various technical studies gives you an idea of what others in the market are likely to do.

9. *Error prevention.* Simple tasks, like inputting prices and calculations, can be accomplished without human error.

This is not a complete list of the advantages. But I hope it gives you an idea of the edge computerized technical analysis can provide.

Computers, like anything else, are nowhere near perfect. You must guard against the false feeling of security they can give you. Never forget what I discussed earlier about price changes being random in nature. Computer analysis often creates the illusion that extremely reliable trends can be uncovered. This often occurs when you are testing or reviewing a trading system using historical data.

Let's say someone is trying to sell you a computer program that analyzes one of the entities underlying an options market, like the futures or stock markets. The seller purports it will alert you to long-term trends and pinpoint prices at which you should enter and exit trades. Running this program using historical data exhibits superior results.

Since the program is using actual price data, what possibly could be wrong? Why couldn't you use the program to spot long-term trends and then buy options to profit from those trends?

When testing or evaluating computer programs using historical data, traders often make some assumptions that are not necessarily reliable when they try to replicate the success in actual trading. Here are some of those hypotheses:

1. *Good fills*—most systems assume you will get good fills. They often mechanically calculate the midrange between when the buy or sell signals are generated. In real life, the fills are not usually as good.

2. *All trades filled*—systems devised by inexperienced or sly traders neglect limit-up or -down days, when no trading takes place. Or days that are illiquid get just as good fills as high-volume days.

3. *Slippage overlooked*—slippage is the difference between the price you expect when you place an order and the price of your actual fill. Slippage is directly related to time and how fast the market is trading. The market, for example, continues to move during (a) the time it takes you to decide whether or not to take a signal the program gives; (b) the time it takes to call your broker; (c) the time it takes him or her to relay your order to the floor; (d) the time the floor clerk needs to relay your order to a floor broker; and (e) the time to read, record, and enter your order. A lot can happen during those five steps, and most computer systems make no allowance for any lost time or market volatility. They get a signal and enter it in the historical trading session. No wonder they get such good fills, so consistently.

4. *Unrealistic commission*—it is not uncommon to review programs that totally neglect the impact of commissions and fees. This can substantially change the hypothetical track record of a system that otherwise looks pretty good.

5. *Human error*—the pressure of trading—making or losing real money, your money—causes errors. I've seen even experienced traders become excited and buy a put when they wanted a call, or order the wrong strike price, or specify the wrong month, etc.

6. *Use of favorable data*—when you decide to test a system, you can select whatever data you wish. If you're evaluating a trend-following system, why not choose a period of time the markets trended? The problem is that when you decide to actually trade, the markets may not be as cooperative.

These problems are so serious that the Commodity Futures Trading Commission (CFTC) requires the following statement to accompany any hypothetical track records used to promote professional traders, called CTAs (commodity trading advisors):

> Hypothetical or simulated performance results have certain inherent limitations. Unlike an actual performance record, simulated results do not represent actual trading. Also, since the trades have not actually been executed, the results may have under- or over-compensated for the impact, if any, of certain market factors, such as lack of liquidity. Simulated trading programs in general are also subject to the fact that they are designed with the benefit of hindsight. No representation is being made that any account will or is likely to achieve profits or losses similar to those shown.

The Securities and Exchange Commission requires a similar statement adjacent to hypothetical performance records of stock portfolio managers used to solicit new accounts.

I'm by no means debunking the use of a computer in options trading. On the contrary, I'm a strong advocate. My point is simply that you must not get caught up in someone else's enthusiasm for a system or software program. If a computerized system could produce superior results (like a marketing system that "always" produces a profit), would it be for sale at any price? Personally, from the success I have seen by some individual traders over the years, there may be some superior programs or systems out there—but I haven't seen any of them for sale.

This takes us back to step 1. If prices are random and there isn't any perfect software, why bother? My answer is the same as before—discipline, organization, consistency, etc. Much of market analysis is very tedious number crunching and record searching. Why not ask your computer to do it for you? Then there is fact finding—primarily prices, statistics, and news. All these tasks are well suited for telecommunication via computers. If all this isn't enough, what about information retrieval? The use of CD-ROM technology and the Internet will take market analysis for individual traders to new levels of intensity.

WHICH SOFTWARE/HARDWARE IS RIGHT FOR YOU?

Before you can select the hardware (CPU, modem, CRT, memory, printer, mouse, etc.), you must know what software you will be using. For example, if you choose a trading system that runs on a Windows 97 operating system, you may not be able to run the program on a UNIX or Macintosh operating platform. Equally important are the attributes—such as speed, memory size, and screen quality—of the system you select. Many of the complex trading programs are memory hogs. They demand a lot of random access memory (RAM) and read-only memory (ROM). Or a high-resolution color monitor might be a requirement to distinguish between the many lines, such as Gann squares or angles, used in the analysis process.

Before you purchase any software, you must decide how you are going to make your trading decisions. How does computerization fit into your trading system? Does it drive your trading, or will you use some software as an auxiliary tool? To aid you in this process, I'll now discuss what I consider the major classes of software. See Appendix 4 for a description of some programs developed specifically for option traders.

BLACK VERSUS WHITE BOX

The first way I classify trading software is whether it is a black or white box. Black box programs are not fully disclosed. They perform a specific function(s), but that's all you know. Additionally, you cannot make any adjustments. You acquire these programs to satisfy a specific need or answer specific questions—you can make little or no changes.

It's common for this type of program to be trendy. It gets "hot," works for a while, and then fades into obscurity. I think the reason is that it handles one type of market, such as a trending one. Once that changes, the program no longer works. It cannot adjust or be adjusted by the user.

Another problem I have with these black box specials is that it is very difficult for someone to design a program that's exactly right for someone else. Therefore, many people buy them only to find them unusable for their system, trading tactics, or personality. Also, certain trading approaches work for one commodity or stock classification, but not all. Which ones are right for a particular software package?

From time to time, I have found a use for certain of these pro-grams. You might want to find one that complements your basic trading system. It could provide you confirmation of what your sys-tem tells you. Let's say your analysis alerts you to a buying oppor-tunity in soybeans. You also have a program that analyzes options. It determines which ones are currently overpriced or underpriced. Additionally, it figures the risk-reward ratio of a variety of strate-gies (spreads, straddles, etc.) and ranks them based on profitability.

Or you may have a program that predicts seasonal patterns. You double-check it to make sure you are not bucking a seasonal trend before you place a trade, or you use this program to alert you to upcoming seasonal opportunities.

White box programs, on the other hand, provide in-depth expla-nations of the analysis, how it works, often including formulas and instructions on adjusting or modifying the output. These programs are flexible—allowing you to adjust to changing markets and your trading style. You drive the software, rather than the software dri-ving you. But these programs require much more of your time than the black box ones.

Naturally, there is much middle ground, the "gray" programs. These permit minor changes and adjustments. I found that one of the keys to distinguishing between black and white is the amount of instruction you get from the program and the amount of documen-tation. The more you learn, the whiter the box. If you get nothing from the program but an answer, it's a black box for sure.

Neither is necessarily right or wrong. It's your needs that count. In some cases, all you want is a reliable answer. In other cases, you may want to learn all there is to know about an analytical technique. Then you want to be able to customize it for your own use.

SINGLE VERSUS MULTIPLE PROGRAMS

Computer programs can do a single function, or they can take the toolbox approach by providing many technical studies. This applies to either black-box- or white-box-type applications.

A typical black toolbox program might, for example, provide price charts. On these charts, you could overlay moving averages, rel-ative strength indexes, Gann lines, etc., by striking a few keys. You would have a very limited ability to make any modifications to what is available. Some of these services can be very extensive in the software programs they provide. They are like supermarkets where you pick and choose whatever you need depending on the situation. This is the type of program that is often available from services that provide live (actual time) quotes.

The quintessential white toolbox program, in my opinion, is CompuTrac. It includes over 65 programs, ranging from relatively simple open-interest studies to very complex ones, like Williams %R or his "Ultimate" Oscillator (see Figure 8–1). The documentation alone could be considered a definitive study on technical analysis.

My last classification includes programs designed to assist you in developing your own trading system. They fall into two basic types. The first is what is known as an expert system, which attempts to mimic a master. You feed your trading rules and experience into the system. It then takes the data and does the analysis. You constantly fine-tune the operation, and it even "learns" on its own by combining the rules or lessons of the expert. These are very sophisticated programs.

A variation of this approach is the neutral network. Neutral-network programs manipulate data with the knowledge of the desired output in hopes of finding the rules or patterns. They require much patience, time, and experience.

Neither the expert nor neutral-network approach, in my opinion, can totally compensate for the randomness of price activity. The rules and patterns governing price activity are constantly in flux. It is doubtful these systems can do more than help you predict the odds on any given trade. But this may be all you are looking for from a system. For a detailed discussion of trading systems, read *The Business ONE Irwin Guide to Trading Systems* by Bruce Babcock, Jr. Which approach suits your needs? Only you can answer that one.

So far the discussion has focused on why you should consider computerizing your trading and the general classes of software available. Now we'll explore a method of selecting the software that suits your needs.

THE IDEAL SYSTEM

There is an approach to problem solving that also works well for selecting software. It's called the "ideal system." What you do is let

FIGURE 8–1

CompuTrac's Programs

Advance-Decline	Moving Average
Alpha-Beta Trend Channel	Normalization
Andrews Pitchfork	Norton High/Low Indicator
Arms Ease of Movement	Notis %V
Average True Range	Oil Crack Spread
Black-Scholes Option Analysis Module	Open Interest
	Oscillator
Bollinger Bands	Parabolic (SAR)
Bolton-Tremblay	Percent Retracement Points and Figure
Commodity Channel Index	
Commodity Selection Index	Planets Module
Compression of Data	Quadrant Level Lines
Crocker Charts	Rate of Change
Cutler RSI	Ratio
Cycle Finder	Relative Strength Index (RSI)
Demand Aggregate	Schultz A/T
Demand Index	Shift
Detrend	Soybean Crush Margin
Directional Movement	Spread
Envelopes	Standard Deviation
Fibonacci Fanlines, Arcs and Time Zones	Stochastic (%K, %D)
	Stoller STARC Bands
Fourier Analysis	Swing Index
Gann Lines	Tirone Level Lines
Gann Square of Nine	TRIX
HAL Momentum	Trading Index
Haurlan Index	Trend Lines
Herrick Payoff Index	Volatility
Linear Regression	Volume
Moving Average Convergence/ Divergence (MACD)	Volume Accumulator
	Wave
McClellan Oscillator	Weighted Close
Median Price	Williams %R
Momentum	Williams "Ultimate" Oscillator

your imagination run wild. You ask yourself: How could I ideally solve this problem? If I had no financial, internal, political, or personal restraints, what would I do? You then write down the ideal solution to your problem.

Once you have brainstormed the situation and outlined the best of all possible approaches, you begin to compromise with reality. What can you afford to do? What can you live without? What do you have time to do? What restraints are real and which are imaginary? How can you circumvent some of the limitations? What's left?

The rationale for this approach is that this exercise often opens new avenues of thought. You find unique and valuable approaches you may have previously overlooked. It's not uncommon that some of these new insights will substantially improve the solution you eventually reach.

To apply the ideal system concept to the selection of software, you prepare a list of all the features you ideally would like to have as part of the software you want. This becomes your checklist as you evaluate potential packages. To keep you in touch with reality, you can mark those that you feel are an absolute necessity. Here's an example:

Software Features Absolutely Necessary (X)
Live price quotations ( )
Chart formations (trend lines, channels, etc.) ( )
Basic studies (RSI, moving averages, etc.) ( )
Advanced studies (Black-Scholes option price calculator) ( )
Unlimited studies (100+ basic and advanced) ( )
Artificial intelligence (expert system) ( )
Neutral network ( )
Historical testing capability and data ( )
Portfolio analysis ( )
Risk-reward ratio calculator ( )
Spreadsheet ( )
News ( )
Weather ( )
Fact database ( )
Order placement ( )
Internet access ( )

There isn't room here, but you could make a shopping list of the hundred plus individual studies, programs, functions, and accessories currently available. Then check the ones you feel fit into your trading system or style of trading.

Your objective is to clarify your thinking. How are you going to select trades? Entry, exit points? What tools do you absolutely need to make these decisions? How much information do you need? How automated are you going to be? Will you input price quotations live? Daily? Manually? Are you going to look to your system for money management assistance, as well as trading? Will you maintain your trading diary on computer as well? Can you communicate your orders to your broker electronically or trade via the Internet?

Once you ascertain what you want or need, reality sets in as you price out your selections. A quote system providing on-line price ticks with a decent variety of studies runs $300 to $500 per month. Additionally, you may have to add something to cover the transmission of the signal (cable, satellite dish, dedicated telephone line, FM antenna, etc.), depending on where you are geographically located. A white toolbox type of program, like CompuTrac 3.21, costs between $695 and $1,895, depending on which modules you acquire. A single study program, like OptionVue, goes for $895.

If you would like to have price quotes but can live with a 10- or 15-minute delay, there are reasonably priced (under $50 per month) alternatives, like DTN and Broadcast Partners' FarmDayTA. The delay allows you to get the quotations without having to pay each exchange a fee, varying from $30 to $50 per month per exchange. Four futures exchanges could cost you $120 to $200 in exchange fees alone each month.

RESEARCH YOUR CHOICE

Let's say you narrow the field to a half-dozen programs that meet your needs. What do you do next?

The first thing I'd recommend is to visit, if possible, someone who is using the system in about the same way you plan to use it. Does it really work the way the promoters claim? Can it provide the information you vitally need to trade successfully?

Next, find out if the software has been battle-tested. Who has traded stock, index, and/or commodity options using the software? Make sure you distinguish between real-time action and simulated or hypothetical trading. For a variety of reasons, most importantly our human emotional responses to pressure, it is not uncommon for historical simulation to look great, while live action falls flat. Some trading approaches appear to be automatic, but in real life trading requires a flow of decisions.

Get a tight fix on two critical inputs—your time and money. Many programs require daily highs, lows, opens, and closes. Some may additionally ask for volume and open interest. How are these data to be entered into your computer? If the answer is electronically, what computer format(s) does the program support? If you have an on-line quote system, will the software accept the data directly or is conversion necessary? If so, is it expensive or even available? The last thing you need is a program you can't update easily or inexpensively.

Always research the user friendliness of the program and customer support supplied by the vendor. Also, check the hours customer support is available compared with when you will normally be using the program.

I always like to have an insight into how many times the program has had major upgrades. This is usually a whole-number change in the version, version 1.0 versus version 2.0. Minor changes often are recorded by a decimal change, version 3.5 versus version 3.6. Beware of version 1.0; it may still have some bugs to be found and corrected. Check to see if you get free upgrades when the software is improved.

Always try to get a demonstration diskette—preferably one that runs the program on some limited basis, as opposed to one that creates a computerized slide show. The actual running demos provide a much better insight into what the program does and "how it feels" to you.

CAVEAT EMPTOR

"Let the buyer beware!" If it sounds too good to be true, it probably is. Therefore, don't order anything that doesn't have a good guarantee. Also, use your credit card, so if you don't like what you get, you can send it back and refuse to pay the charge.

There really are some excellent, well-written programs available. It takes time and research to find them.

There are other positive-negative aspects to the influence of the computer chip on the markets we trade, particularly for the underlying markets. For example, the futures markets and the options-on-futures can be extremely volatile. At times, they can be characterized as nervous. This characteristic makes them very responsive to news and herd psychology.

In B.C.C. (before computer chips) times, someone physically close to the markets would become aware of market-making news and trade it. The word would spread out from the pits. Late on day 1 or on day 2, brokers would be telling their customers about it. By day 3, it would be on radio and television.

The closer you were to the floor, the sooner you got in the market on the right side to take advantage of the impact of news. The farther from the epicenter, the more of the move you missed. In those times, professional traders had a strategic advantage. The periodontist in Peoria got the word too late—just in time to get creamed by the blow-off top.

The good news is that a trader in Jackson Hole now has cable, a satellite dish, Internet access, and a personal computer—she or he is wired to the world! These traders get the word just as fast as, or often faster than, the scalpers in the pits. The playing field has leveled somewhat. More important are the studies and analyses that can be done by those with the computer horsepower.

The trading edge has shifted from the floor traders to the professional traders. The former chairman of the Chicago Board of Trade, William F. O'Connor, described this phenomenon in this way in the *Chicago Tribune:* "Exchange members used to have a time and place advantage. But the edge is no longer. Simply put, the information advantage has gone upstairs." The ones with the biggest advantage now appear to be those trading massive portfolios for institutions or funds. What's happening in the futures markets mirrors the impact of program trading that took hold of the stock market over the last decade.

The more chip power you can put behind your order tickets, the more responsive your trading can be. Or at least that's the current theory. There are systems like the Market Information Machine (MIM) created by Logical Information Machine, Inc., of Austin, Texas. It's nothing more than a massive database of the stock and futures markets. What makes it unique is what is known as declarative programming paradigms. This feature allows the trader to instantly access all the data in the database, which are all stored in the computer main memory. Normally, data are housed on disks. When presented with a query, the computer searches disk after disk, or drive after drive, to find the answer. It can take from minutes, to hours, to days. MIM queries go straight to the main memory, and the answers come "instantaneously," regardless of the complexity of the question.

MIM is particularly useful for traders seeking correlations between two or more market events. You might ask MIM what to expect from the S&P 500 when the consumer price index is down 5 percent over the last 6 months and the S&P 500's price-earnings ratio is down an equal amount. You could ask for a weekly price chart for the 10 trading days following each time this event occurred. MIM delivers these charts in seconds.

But don't get your hopes up, since the cost is prohibitive for most individual traders. You'd need a $15,000 computer workstation and approximately $3,000 a month to access the data. You may find a brokerage firm that has one though.

Computer analysis has also changed the way the markets perform. Or at least that's my perception. Over the last decade, I have seen more and more of what I characterize as false technical signals or false breakouts in the futures markets. I attribute this to the ever-growing number of traders relying on computerized trend-following systems. These traders all get a signal at about the same time. Their collective responses make the signal a self-fulfilling prophecy. But few of these analysts are truly committed to the trade. After it moves a little bit in their favor, they get cold feet and exit the trade. Their exit produces a reversal signal and the phenomenon repeats itself.

This brings up another important insight into the markets you should be aware of. It's called the "pot never boils when you watch it syndrome." If a market move takes a long time to occur, for example, and the whole trading world is waiting for a trend line to be broken, the break usually becomes a nonevent. In other words, closely watched and well-publicized technical formations usually fizzle out. I remember a well-publicized head and shoulder in the corn market, written about in at least six advisory newsletters—it never reached its potential. Too much expectation was built into the market. If you know something is going to happen and you know everyone else also knows, you can't become really committed to it.

If, on the other hand, you think you know a secret (I'm not referring to insider information, which is illegal to trade)—something is going to cause a stock to make a mega move, and you're the "only" one with that information—you are more likely to stick with the trade. The less press, the less explanation available about an upcoming move, the more likely it will be a big trade.

The computer chip, which makes communications faster and analysis of large amounts of data easier for larger numbers of traders, has forever changed the way markets move. But there are still plenty of opportunities. Actually, there are more opportunities now for small traders than ever before.

C H A P T E R   9

Selecting the Broker That's Right for You

"Because I don't trust him, we are friends."

Bertolt Brecht
(Mother Courage, *1939*)

Key Concepts

- ◆ Understanding the Basic Conflicts of Interests
- ◆ Matching Your Needs to Your Broker's Resources
- ◆ The Art of Managing Your Broker
- ◆ Choosing between a Full-Service and Discount Brokerage Firm

There is a basic conflict of interest between buyers and sellers. Your stock or commodity broker is the seller; you, naturally, are the buyer.

This conflict of interest is not unique to the investment industry. Brokers are middlemen. Their function is to bring buyers and sellers together. Rarely do they get paid by both parties or a disinterested third party. Normally, the seller pays the sales commission. This means the broker works for the seller and has the seller's interest utmost in mind.

For example, when you buy a home or a car, doesn't the real estate agent or the salesperson receive a portion of what you pay as his or her compensation? Yet we all buy homes, cars, insurance, and

hundreds of other things sold by some sort of intermediary in the normal course of our everyday business.

Most brokers in the investment arena work on straight commissions, or mostly commissions. This means that if they do not sell
you anything, they do not get paid. Even brokers on salary, a new
wrinkle among discounters, must make sales to keep their jobs and
receive raises.

These are serious conflicts of interest you need to be aware of—
but they are nothing unusual, nor should they be disturbing. You
just need to be sensitive to them. Brokers are told that while their
direct interest may be somewhat different from that of their customers, their goals can be the same. Brokers know if they help their
clients make money, they will make money. Setting mutual goals
helps deal with some of these situations.

As you read this chapter, you should think about the preceding
chapters. A successful broker-client relation is a partnership. Each
party contributes substantially to the success of the trading. You seek
a broker who complements your specific need for market knowledge, experience, background, trading expertise, equipment, research,
etc. Therefore, as you work your way through the broker selection
process, you create a list of what you want from a broker. Start with
a list for an *ideal broker* and then work your way back to reality.

Let's very quickly review what we've covered so far. We'll just
run through each chapter and list some of what you might ask yourself to define what the ideal broker for you would be.

Chapter 1. What do you need from a broker in the way of
basic knowledge and background? About trading options?
About markets?

Chapter 2 & 3. What help do you need regarding understanding and selecting strategies?

Chapter 4. How will you reach an opinion of market direction
and strength? Fundamental or technical analysis? What research, facts, information, etc., will you seek from your broker?

Chapter 5. Will you expect volatility estimates and insights
from your broker?

Chapter 6. How does your broker fit into your trading plan?

Chapter 7. What help do you need putting your trading system together?

Chapter 8. Will you be looking to your broker for price quotations, theoretical prices, implied volatility, or other computer-generated trading signals?

Another area many traders look to their brokers for is feedback from the trading pits. Is it important for your broker to know what's going on in Chicago, New York, wherever? Should your broker have actual experience on the trading floor? You may, for example, be approached by a broker who has traded for his own account in a pit. He'll tell you this is extremely valuable to you. Is it so?

One way to approach this question is to compare on-floor with off-floor trading. Let's talk first about the on-floor trader. There are basically two types. The first is "locals" trading their own account; the second is brokers filling other people's orders. Additionally, some exchanges allow brokers to dual-trade—for their own account and off-floor customers.

Probably the most prominent characteristic of on-floor, local traders is their short-term view of the markets. Rarely do these type of traders hold positions overnight. These guys are often scalpers, trying to shave a cent, or even a fraction of a cent, in order to profit on the contracts they trade. Since they are usually members of the exchange, their per-transaction costs are extremely low, while their volume is extremely high.

Scalpers provide a very important function—they are the price police and market makers. Their attention is constantly directed at the bid-ask range, looking for anything that's overpriced or underpriced. If they spot anything out of line, they jump on it and hold it until it is fairly priced again. They make their profits on the correction. The market benefit is that bid and ask prices are kept in fair ranges.

Their activity also adds liquidity to the markets. This is critical to you as an off-floor trader for two reasons. First, it allows you to enter or exit a market promptly. Second, the range between the bid and the ask price is tighter, which usually means you get better fills.

Liquidity is a particularly important issue in the options market because it is often lacking in some strike prices—particularly those out-of-the-money. Second, in some of these illiquid markets, brokers are "not held" for all their orders. This means that if an option order is entered in the market within the bid-ask range, it still may not be filled, if another broker doesn't have an offsetting order for it. In other words, the brokers do not have to "make a market" for all strike prices.

Last of all, when an order comes into the pits at the upper limit or even outside the bid-ask range, the market may take that order as the new bid or ask price. In other words, the range stretches in these illiquid markets to capture any order it can. This sometimes results in a bad fill for the retail (off-floor) customer. Before your (off-floor) broker places an order for you, he or she may want to call the pit to get a feel for the current liquidity.

The other type of floor broker is one working for a broker-dealer or a futures commission merchant (FCM) filling orders for their customers. These traders learn the internal workings of the exchange, but may not be as shortsighted as the scalper.

The importance to you of knowing how the pits really work—and the reason good brokers cultivate friendships with floor personnel—is that this knowledge can help get better fills. Or the floor brokers can help recommend where to place stops. The people on the floor are constantly talking to each other. Ex-floor traders can read interday pivot points better, due to their understanding of the day trader's mentality. They want to know who's behind a bid or ask order. Is it a giant institutional trader or a small speculator? You can't tell no matter how hard or how long you stare at a price quote screen.

The floor people also know who is holding orders and at what price. Are there some stop orders at a certain price that will stop or slow down an uptrend or downtrend? Information like this can only come from the floor.

Therefore, it is not so important that your broker has actually braved the cruel world of the pits. It is critical that he or she, or the firm you trade through, have contacts to get reliable information fast. As a matter of fact, if a prospective broker tells you he or she traded as a local, this doesn't guarantee he or she is a better broker. Floor traders, particularly scalpers, usually trade only a few markets. They may not be well rounded in all the stocks and/or commodities you wish to trade.

When trading that single market, floor traders are not looking at the medium to long term. They are not even looking at the short term—a few days—as most off-floor traders classify it. Overcoming this perspective can be very difficult. Ex-scalpers may encourage customers to take profits too soon. They can leave a lot of money in the pits. Also, since they traded only a few markets, they may lack an understanding of the importance of diversifying your portfolio.

My point is simply this—information from the pits can be very important at certain times. Your broker needs the capability of delivering it to you when necessary. A good broker is one that satisfies your needs; it doesn't matter if he or she ever stepped foot in a trading pit. Your needs are the critical issue.

HOW DO YOU SPOT A "GOOD" BROKER?

First off, the key word is "good." It means good for you, your trading style, your personality, and your needs. Do you need assistance in selecting trades, tracking them, managing money, getting technical and fundamental information, placing orders, obtaining exchange floor reports, etc.?

One of the best places to develop a list of what you want from a broker is to develop a list of your limitations. Your broker should become your partner; you make money together. Your broker must earn his or her commissions.

There are at least four common limitations most individual traders must overcome. The first is time—time to follow the markets and to sort out the important information from the false rumors that clutter the pits. The second is the flow and cost of information available about the markets. As fast moving as the futures markets are, you often need tick-by-tick price quotation equipment. At times, the stock market is equally fluid. Additionally, you need immediate access to the wire services and other forms of news. Therefore, you may require some expensive and sophisticated equipment, which most individuals cannot justify as part-time traders. The Internet has eased the cost of some of this, but to have access to live price quotations you still have to pay substantial monthly fees to the exchanges.

Next is the experience that only comes from trading the markets day after day, year after year, facing bull markets and bear rallies, spotting the major moves early, and avoiding getting whipsawed by false breakouts.

Last of all, there is just an enormous amount of background information you need at your fingertips to trade and to avoid making expensive errors. You need to know all the specifications of the options you're trading, the hours the markets are open, what days they are closed, if they are closing early for some reason, how to place orders, when important government reports that will impact markets are to be released, what rumors are sweeping the ex-changes—and it goes on and on.

Therefore, develop your list of what you are going to look for in a broker before you begin to interview (usually by telephone) broker candidates. And, I do mean candidates. Use all the skills you learned over the years in hiring employees or vendors, selecting schools for yourself or your children, and making other critical buying decisions. Make a "T" list on all candidates after you talk with them (see to Figure 9–1).

If you do this for each broker you interview, the selection process becomes much easier. From my experience with working with and training hundreds of brokers over the years, I have found that the following are some characteristics common to "good" brokers:

- *Availability.* Good brokers are available to their customers whenever they are needed. This often equates to putting in long hours in order to cover everything from the early grain markets to the evening Treasury bond sessions, or to track and trade the stock markets from New York to the Pacific coast to Japan to Europe and back. When they aren't personally available, they see to it that an assistant or associate is. You're never out of touch. Ask prospective brokers about how "reachable" they are. Then, when you open an account, test them (call them at 8:00 a.m. and again at 8:00 p.m.) before a market emergency occurs. Do they offer you their home phone number?

- *Responsiveness.* You need a broker who'll help you overcome your limitations and satisfy your needs. Therefore, this person must be responsive. In the interview process, ask him or her some obscure piece of information such as when the Federal Reserve is releasing the next Beige Book. Find out in advance how the broker performs. Ask how fast you'll get fills. Try to discern how hard the broker will push to respond to your needs.

- *Honesty.* Naturally, you must trust and have confidence in your broker. You can often learn a lot about sincerity when you ask about the risk of futures trading or a specific stock trade. If the broker evades or glosses over the high risk involved, beware. You should receive a serious, no-holds-

F I G U R E   9–1

Broker Interview "T" List

Broker's Name: _____

Firm: _____

Phone Number: (_____)_____

My Need and/or Limitations *Pro/Con*

	Pro	Con
1. Is he or she a stock (commodity) options broker?	Commodity	
2. Is he or she an experienced options trader?	3 years	
3. Appears knowledgeable about my futures markets (stocks)?		No, but other ones
4. Live Quotes	Yes	
5. Black-Scholes Model available	Yes	
6. Sounds like he has good trading system		Not sure?
7. How does firm sound?	Okay	
8. Commission rates?		$125 (all up front)
9. Good floor contacts?	Yes	

My overall impression is:

_____

_____

_____

_____

barred answer from prospective brokers regarding the buying of options—the risk is limited to the cost of the option (premium) and the transaction costs (brokerage commission and fee), but you can lose 100 percent of this amount; most options expire worthless; few traders make money in options. If the broker tells you options trading is a serious investment with high risks, it's a big plus.

◆ *Dealing with Losses.* Under the stress of losing other people's money and the pressures of trading, many brokers come apart. The first sign is often a lack of communication with the losing client. This is almost impossible to detect in the early interview stage of the selection process; but when you start to trade, you should become sensitive to it. Also ask for some names of past customers you can call. Contact these customers and ask them how the broker performed when the account was in a drawdown. The best brokers don't let it bother them because losing money is a large part of options trading. Don't get me wrong. They take it seriously, but they can manage it and go on.

◆ *Trading.* Find out how they trade. What system do they use? Are they fundamentalists, technicians, or a combination? Have them "sell you a trade." Just ask them what their current trade recommendation is. You want to learn how well they present an opportunity to you. Is it well thought out? Convincing? Have they done their homework?

◆ *Discipline.* Good traders and brokers are very disciplined. They spend hours studying the market and refining their system. Weaker brokers will try to emotionally sell you the "hot trade of the day." Probe to get a feel for the daily routine of brokers you are interviewing. Do they keep a trading diary? What do they do each day that assures that your positions will be reviewed?

◆ *Experience, Knowledge, Track Record.* If you know more or have traded more than a prospective broker, who's going to help whom? Find out what they have done, what they know, and how that's going to make you more successful.

The question of a trading track record is tricky. Most brokers do not have one. If they do not take total discretion in an account (limited power of attorney that permits them to actually trade the account as if it were their own), the customer must approve each trade. Therefore, the broker is merely a consultant and the customer is the trader. The broker cannot use these accounts to create a track record.

Next, there are some very stringent regulations imposed by the Securities and Exchange and the Commodity Futures Trading Commissions regarding track records. These regulations make it difficult and expensive for a broker to prepare a track record. This is another reason most brokers do not have one. Personally, if a broker shows or offers a track record that is not part of a full-blown disclosure document that has been reviewed by the federal regulators, you should be very suspicious.

These are just a few of the areas to which you must be sensitive as you select and work with a broker. Always remember that it is your money that's being invested in the market; you're the one who is financially responsible.

As you interview prospects, you'll find two basic types—discounters and full-service brokers. You'll also find some who claim to provide full service at discount prices. All too often, new traders make their decision based on the prices of the service, as opposed to their needs. Before you talk to any broker, prepare a written list of what you want—need—from your broker. Don't get sucked into the least expensive service if it can't deliver what it takes to make your trading successful.

Let's talk dollars for a moment. The per-trade cost at a discounter ranges between $20 and $35 for futures. Stocks depend on volume, usually with a minimum of $35 to $40 for the first trade. The cost of electronic or Internet trading will be lower, but may not include regular access to a broker. In most cases, the trader places toll-free telephone calls to the order desk at the discounter's headquarters. An order clerk takes the order and relays it to the order desk at the appropriate exchange.

To utilize this type of service, you need a lot of experience. The order clerk is not in a position to assist you in choosing or evaluating a trade. The order clerk's function is strictly to execute orders. In most cases, he or she will not even provide current price quotes or other information—like highs, lows, trading ranges, trends, market news, volatility, etc.—that you may want at the time you are placing your order. Discount costs match the services, as a general rule. Discounters serve full-time traders, who have the time and money to generate all the information they need for their trading system.

Another variation in discounting is direct access to the futures or stock markets via personal computers/Internet. Most firms limit stock options to covered options. There are two routes you can take. One is to use public database services, such as AOL, CompuServe, or Prodigy, and the other is to connect to the mainframe computer of the brokerage firm. Direct access earns an additional 10 percent discount in most cases. The main advantage is being able to trade whenever the spirit inspires you.

Full-service firms are set up to work in tandem with both experienced and inexperienced traders who need assistance in gathering and interpreting information. You would expect a full-service broker to assist with your understanding of the market and supply information (news, price quotes, theoretical option prices, volatility estimates, etc.) and research. You look to these firms for ideas, advice, and experience. You want to find a firm that can truly become your partner.

The cost per trade for full service is naturally higher. It ranges from $50 to over $100 per trade for futures. For stocks, options are usually sold in dollar packages. The broker will offer you X number of options on a certain stock for $2,500 with the commission built into this figure. Another consideration is whether the brokerage firm charges commissions on options on a half-in, half-out basis—half the commission when you place the trade and half when (and if) you offset. Many options expire worthless. Therefore, if you let an option expire, you would not be charged the half-out side of the commission.

There is an old saying in the industry—"If the client is making money, the commission rate isn't an issue." There is a lot of truth to this axiom. Your objective should be to make a profit on your investment. If you don't get what you need from your broker, it doesn't matter what you pay.

CHAPTER 10

Knowing a Little about Federal Regulations Can Sometimes Smooth Rough Waters

"Money doesn't talk, it swears."

Bob Dylan (1941–)

Key Concepts

- Licensing of Stock and Futures Brokers
- Understanding Exactly What the Account Documents Are Telling You
- Learn Your Rights as an Investor
- What Is and What Isn't a Legitimate Complaint against a Broker
- Reading Daily and Monthly Brokerage Statements
- Type of Trading Discretion You Can Give Your Broker

People selling investments and services may be licensed or unlicensed. These licenses are granted on state and/or federal levels. Knowing the supervisory structure under which your options broker is licensed provides you with some insights into your recourse if you ever feel you have been unfairly treated.

Let's start with the option brokers who are not required to be licensed. They usually sell what are known as dealer options, normally gold or silver bullion. They sell options to buy on the cash metals markets. Here you are buying the actual metal, when and if you exercise your option. The prices of these options and the brokerage commissions traditionally cost more than options on futures or stocks, but these options may be for longer periods of time.

Since these brokers are unlicensed, there is no regulatory association or commission to appeal to if you have an unresolved complaint with your broker or his or her firm. Your recourse would be to the Better Business Bureau, your state's attorney general, the FBI, the Fraud Division of the United States Postal Service, or local police authorities. While all of these organizations do a fine job, none specialize in the particular problem of dealer options. Unless the problem is widespread, you may not get the attention you deserve. Of course, you can always bring a civil or criminal suit directly via your attorney. This can be expensive and is not a reasonable alternative unless a large sum of money is in question. Plus many of these people are transient—you may not be able to get a hold of them when you realize they've misrepresented the deal they sold you.

Is the problem of customer complaints dealt with any better when the broker is licensed? First, the licensing process can be on the state or federal level. Doctors, lawyers, and real estate agents are supervised and licensed on a state-by-state basis. In most cases, there is no reciprocity between states. For example, if you are registered in one state, you must still go through the registration process required by each state in which you wish to conduct business. This usually requires passing a competency test. In some cases, registration is available only to residents of that state.

When it comes to selling exchange-traded options on the futures markets, licensing is done nationally. Stock option brokers are required to pass a national exam. For futures or commodity option brokers, their license is nationwide. Stockbrokers must additionally register in each state they do business, but need only take the test once. As was mentioned when I was discussing the differences between options on stock and futures, these different types of options are generally traded on different exchanges, stock exchanges for stock options and futures exchanges for options-on-futures.

The key for broker licensing is the exchange. To accept orders for a futures exchange, you must be a licensed futures broker. This requires passing the National Commodities Futures Exam, known as the Series 3 Exam. Conversely, to take orders from customers for stock options, you must be a registered stockbroker. This requires passing the General Securities Exam, known as the Series 7 Exam. Series 3 and 7 are just part of the overall licensing possibilities for someone to sell various investment products. For example, one would need to pass the Series 6 Exam, to just sell mutual funds, and the Series 62 to sell corporate securities. The possibilities go on and on. There is no prohibition against a broker being registered as both a Series 3 and 7, but it is somewhat unusual. It's even more unusual, as noted previously, for a broker to trade both stock and futures options. The problem is being able to stay on top of these two very diverse markets.

Conducting business with individual brokers and brokerage firms requiring registration means there are regulators, who understand options trading, monitoring sales and trading activities. This is not a guarantee that a regulated investment will be a better investment than a nonregulated one. Profitability is a function of savvy investing, not regulations. Regulators do not always concern themselves with the profitability of specific investments or trades. Their raison d'être is to enforce the rules.

If regulation doesn't protect you against losses, what does it do? First, every person registered to sell exchange-traded options passes either the Series 3 or 7 Exam. Therefore, you know you are talking to someone who at least knows the basics.

Second, the regulatory organizations, the National Security Dealers Association (NASD) for stockbrokers and National Futures Association (NFA) for options-on-futures, monitor the activities of the members. The organizations do this by investigating any customer complaints they receive and by making periodic visits to members' offices, called compliance audits. At these visits, the regulators audit the member firms for compliance with all their regulations. These regulations cover back office (accounting practices, order placement, customer reporting) operations, as well as sales practices.

One of the important areas that NASD and NFA compliance auditors check is the account papers you will be required to fill out in order to open an option trading account. Account papers are the basic agreement between you and the brokerage firm (FCM).

You need to read them carefully because they detail:

1. The risk you face trading options
2. Your responsibilities

3. Your rights (and the rights you waive) as a customer of a specific brokerage firm
4. The services you'll receive
5. The cost of the brokerage services
6. The rights the brokerage firm assumes when you sign the account papers

Think about account papers from three points of view. First of all, these forms are written by attorneys working for the brokerage firms. Therefore, it is only common sense to assume that much of the language and stipulations will be for the protection of the brokerage firms. For example, by signing account papers you give your broker the right to close out your positions, without prior notice, if your account does not have the proper funds in it.

In its self-interest and for its self-preservation, a brokerage firm will do all in its power to collect what it is due. If a client owes margin money to a brokerage firm, the firm must cover the trade, or the exchange on which it is trading will take action. The exchange and its clearing corporation will not allow a brokerage firm to hold positions that are not properly margined. The integrity of the market must be preserved, so the trader on the other side of your trade can immediately act. It is for this reason that the entire industry is so concerned about every trade being properly financed.

If you receive a margin call or a call for more funds, you may have to wire them within 24 hours or lose your position(s). This is another condition you agree to when you sign the account papers.

Another point of view that comes out loud and clear in the account forms is that of the regulators. The Securities and Exchange Commission and the Commodity Futures Trading Commission oversee the NASD and the NFA, respectively. The SEC and CFTC receive their authority from Congress. (See Figure 10–1.)

These two regulators want to be sure you understand the risks and that you are qualified to trade. The risks of loss of all your trading equity is spelled out in frank terms. After reading it, I sometimes wonder why anyone trades—but the glimmer of superior returns to equity seems to prevail. Also, as many S&L shareholders and commercial real estate developers of the early 1900s will attest, there are no foolproof investments—no matter how much regulation exists!

F I G U R E   10–1

Oversight Hierarchy

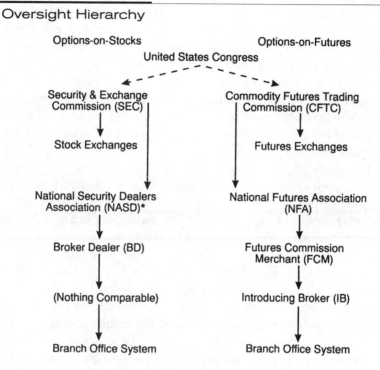

*Stock brokers must also register in each state in which they do business.

The federal regulatory procedures require stock or commodity futures brokers to pass competency exams. Additionally, the business practices of the brokers and their firms are supervised to assure ethical standards are maintained. The NASD and NFA also provide a formal customer complaint procedure.

To determine your suitability to trade options and to satisfy the regulators' dictum that brokers must know their customers, the account papers include a series of questions you are required to answer. Specifically, you will be asked:

- Your true name and address
- Your principal occupation
- Your current estimated annual income and net worth
- Your age
- Your previous investing experience

If, for some reason, you prefer not to answer one or more of these questions, say the one on net worth, you'll be asked to put your reasons in writing. And you may be asked to acknowledge you have a net worth equivalent to at least the minimum acceptable to the brokerage firm. That might be something like $100,000, for example. If your reason for not answering a question was acceptable, your account may be opened. On the other hand, brokerage firms do not open every account presented to them by their brokers. They have the right to refuse your account, and they will if they cannot determine your suitability.

The third point of view to consider as you study the account papers is your own. Do you understand and are you prepared to live up to all the conditions? For example, if you were to write an uncovered option or exercise an option for a position in the futures market, you now have "unlimited" risk of loss. You can lose more than what you have invested in the original position and you are subject to "on-demand" margin calls. This is completely different from buying an option with its predefined risk.

Account papers are not written to protect the investor. They do let you know exactly what your risks and rights are—if you read them!

There is one particular form that is part of all account papers that isn't always explained in as much detail as it probably should be. It is the "Predispute Arbitration Agreement." It binds you and your brokerage firm, including all its employees and owners, to final arbitration. This agreement usually includes the following additional stipulations:

- Both parties waive their rights to seek remedies in a court of law. In other words, you can't sue your broker.
- The discovery or investigative portion of the hearings is more limited than it would be in a court of law. You don't have to spend a lot of time and money digging up information.
- The arbitrator's awards or decisions are not required to include factual findings or legal reasoning. This means the individual or panel judging the case does not have to stick strictly to the letter of the law. If those judging think someone has been wronged, but no law has been violated, they can still find for the complainant.

- The avenues available to seek modifications of rulings are limited. In other words, you waive some of your rights to appeal. You agreed to binding arbitration.
- The panel of arbitrators normally will be composed mostly of people affiliated with the futures or security industry. The reason this occurs is that it is often difficult to find unbiased arbitrators who also have a solid understanding of options. This is not necessarily negative to the individual bringing the complaint. Industry insiders are often very stern with their compatriots if they believe the actions of those compatriots are hurting the industry. In most cases, you can request a "mixed" panel of industry and nonindustry people.

The specific arbitrators or arbitration panels depend on whether the complaint is against a stock or commodity brokerage firm. Stock companies tend to use arbitrators from the exchange on which the trade took place, the NASD, or the American Arbitration Association. With futures firms, the National Futures Association handles all arbitration.

Geography also can come into play. The complainant isn't expected to travel a long distance. If you had a complaint about a trade on the New York Stock Exchange and you lived in Sandusky, Ohio, you wouldn't be expected to travel to New York City. Other arrangements would be made. Occasionally, the arbitration takes place by phone, for example, or in the nearest major city.

The objective is to obtain fast, inexpensive, and fair resolutions to disputes. You don't need a lawyer, nor do you need to be well versed in rules of evidence. It often comes down to your word against your broker's.

The $64 question is: Should you sign the "Predispute Arbitration Agreement"? By not signing it, you do not give up the right to ask for arbitration. Additionally, the NASD and the NFA can compel members to respond to the request for arbitration of clients. If you are a small trader with a small account, it is probably to your advantage, in my opinion, to sign it. Some firms, like the discount stockbrokers, have a predispute clause built into their customer agreement.

When or if you have a complaint with your broker or his or her firm, you contact the NASD for stock problems and the NFA for futures. The NASD will give you the name of a staff person, who guides you through the process. The first thing this person asks for is a written summation. After reviewing this, the staff member will assist you.

For disputes with futures brokers, you contact the NFA. The NFA will first attempt to arbitrate it. If this doesn't resolve it, you file a formal complaint and a hearing is set. You might want to call the NFA for a copy of its booklet *A Guide to Arbitration of Customer Disputes*. Or you can file a complaint with the CFTC. If you are considering this approach, contact the CFTC for a copy of its free booklet *Questions & Answers about How You Can Resolve a Commodity Market-Related Dispute*. The CFTC has basically three reparation procedures:

- Voluntary
- Summary
- Formal

The voluntary procedure is used for claims under $10,000. It is administered by judgment officers appointed by the CFTC. Both parties have an opportunity to uncover facts ("discovery") and present their arguments in writing. There is no oral hearing. Decisions are final, and no appeal is permitted. A nonrefundable $25 fee is required.

The summary procedure is similar to the voluntary one, but it allows for both a limited oral presentation and a written presentation, which takes place in Washington, D.C. Further, you can appeal the decision to the CFTC and to a court of law if you are still not satisfied. It handles complaints of $10,000 and less and requires a $50 nonrefundable filing fee.

The formal procedure is designed to handle major complaints, over $10,000. A courtlike hearing is conducted in one of twenty locations throughout the United States before an administrative judge. You can be represented by an attorney if you wish. Appeals to the CFTC and the courts are possible. A $200 nonrefundable fee is required to file.

Besides the NFA and the CFTC, you can take a complaint to the American Arbitration Association or file a civil suit. If you think your broker or his or her firm has committed a criminal offense, you can contact the Federal Bureau of Investigation; or if the U.S. mail (your monthly account statement) was involved, the one to contact is the chief postal inspector of the United States Postal Service.

If you are invested in a limited partnership or a fund, you may be able to appeal to your local state attorney general or the Securities and Exchange Commission. Other places with which you may check are the Better Business Bureau and the Federal Trade Commission.

For help deciding your most effective alternative, you should probably sit down and discuss it with your attorney. Please keep in mind two very critical considerations:

1. Your most effective protection results from systematically conducting due diligence research when selecting a broker. Act with reason—don't get caught up in an emotional response to the potential the stock or futures markets may offer.

2. In my opinion, sticking to investments that are federally regulated usually gives you a better dispute resolution process than unregulated investments.

Over the many years I was in the securities business, the most beneficial arbitration was done between brokers and clients. Once it goes beyond this stage, the costs and complications seem to outweigh the results. My advice is to work hard at this level to keep the lines of communication open. Efforts in this area pay dividends in better trading, fewer errors, and more enjoyable trading.

You may be wondering what can be considered as grounds for a legitimate complaint. Losing money? Bad advice? A trade gone sour? An honest error? None of the above. All these should be expected. Anyone who has ever traded options, or just about anything else for that matter, will attest that losing money on some trades is a way of life. Bad advice simply means your broker, or his or her company's research, cannot foretell the future accurately. Nobody can. This shouldn't be a surprise either. Nor should an occasional error upset you, especially when you consider the enormous trading volume that takes place.

The following are some of the grounds for a formal complaint:

- Being high-pressured into opening an account or taking a recommended trade
- Receiving unreasonable promises, like—"You can't lose money trading options!"
- Any fraudulent or deceitful communications made to you by your broker or his or her firm

- Excessive trading in an account that a third party is directing for you
- Trading your account without your permission or direction
- Uncorrected errors made in your account

In just about all these situations, the broker is being untruthful. To high-pressure you into a trade, for example, he or she may exaggerate the profit potential or the urgency with which you must act.

The last three list items describe problems you can monitor, if you pay close attention to the statements you receive from your broker. Each transaction you make—a trade, deposit, withdrawal, exercise of an option, etc.—generates a statement. Additionally, each month you'll receive a monthly summation of the month's activities.

Because these written reports arrive two to three days after a trade has taken place, all too many option traders don't pay any attention to them. You need to study these statements as closely as you balance your checkbook, maybe even closer. Ask yourself:

- Are the commissions and fees correct?
- Do I remember making all these trades? (Always keep at least a log of trades, if not a diary.)
- Are the entry and exit prices the same as the fills my broker gave me?
- Am I on the correct (put or call) side of each trade?
- Have proper debits and credits been made?
- Is the extension from a per unit to a full-contract price correct?
- Are my name, address, and account number correct?

If you question anything at all, call your broker promptly. Don't wait. The longer you put it off, the harder it is for your broker to reconstruct the situation and make sense out of it. Good brokers are extremely busy but are always ready to help you (see Figure 10–2).

DISCRETIONARY ACCOUNTS

Some brokers are so helpful, they'll even trade for you. This is called "taking trading discretion" in an account, and there are varying degrees of discretion.

F I G U R E   10–2

Combined Commodity Statement

COMBINED COMMODITY STATEMENT

MR. I. M. WELLOFF
1600 FAIRVIEW BLVD.
PROSPEROUS, PA 19013

DATE	BOUGHT LONG	SOLD SHORT	COMMODITY/OPTION DESCRIPTION	PRICE	AMOUNT DEBIT	AMOUNT CREDIT
	SEGREGATED FUNDS					
05/11/98	ACCOUNT BALANCE		SEGREGATED FUNDS			5,100.90
			----------CONFIRMATION----------			
WE HAVE MADE THIS DAY THE FOLLOWING TRADES FOR YOUR ACCOUNT AND RISK						
	1		DEC 93 CORN C 250	.10	500.00	
			PREMIUM PAID		500.00	
	1*		OPTION EXPIRATION 11/20/93	COMM	75.00	
				FEE	4.25	
			TOTAL COMMISSIONS & FEES		79.25	
			NET PROFIT OR LOSS FROM FUTURES			00
			NET PREMIUMS PAID/RECEIVED FOR OPTIONS		500.00	
CURRENT ACCOUNT BALANCE			SEGREGATED FUNDS			4,521.65
			----------OPEN POSITIONS----------			
03/10/98	2		JUL 93 COMEX SILVER C 600	.25		3,000.00
			PREMIUM VALUE			3,000.00
			SETTLEMENT PRICE	.30		
	1		SEPT 93 SOYBEAN P 500	.12		500.00
			PREMIUM VALUE			500.00
			SETTLEMENT PRICE	.10		
	1		DEC 93 CORN C 250	.10		400.00
			PREMIUM VALUE			400.00
			SETTLEMENT PRICE	.08		
			TOTAL OPEN TRADE EQUITY		.00	
			TOTAL EQUITY			4,521.65
			TOTAL LONG OPTION MARKET VALUE			3,900.00
			TOTAL SHORT OPTION MARKET VALUE		.00	
			NET LIQUIDATING VALUE			8,421.65
			----------SECURITIES ON DEPOSIT----------		----VALUE----	
06/07/98		10,000	US TREASURY BILL			9,000.00
			TOTAL			9,000.00

Grains in 000's   Retain for Tax Records.   Terms and Conditions on Reverse Side.

You'll receive a statement each time some activity occurs in your trading account. This could be an addition, withdrawal, or trade. Monthly you receive a statement that summarizes the month's activity. Always carefully check your statements for accuracy.

On the lowest level, you give your broker either time or price discretion. This means you have decided which stock or commodity you wish to buy or sell, the side of the market (put or call), and the quantity. You give your broker instructions to act at a certain time or when a price is hit.

The next notch up on the discretionary ladder is the guided account. Here a professional trader suggests specific trades. Your only choice is either to accept or to reject the trades. If you accept, the trade is executed. Unlike the normal relationship with a broker, there is no discussion or meeting of the minds. Either you accept or you pass on the recommendations.

With the next level of discretion, you sign a limited power of attorney giving your broker authority to trade your account as he or she sees fit. If the person actually trading the account is someone other than your broker, called a third-party advisor, the advisor may be required to supply you with a disclosure document. This is called a managed account. A disclosure document is basically a prospectus explaining all the details of the advisor's past trading, risk factors, strategies, and personal history. Generally, advisors who qualify as exceptions to the disclosure document requirement are an immediate family member of yours, an advisor who does not solicit accounts and manages only a few accounts each year, and a properly licensed broker with 2 years experience as a broker.

Another possibility is a fund, pool, or limited partnership. This could be public or private. In either case, you receive a formal placement memo or prospectus, which provides all the details. It would be rare for a fund or a pool to only trade options. Most trade stocks or commodities primarily, with the right to trade options when special opportunities present themselves.

CHAPTER 11

The Single Biggest Mistake New Traders Make—Plus a Few Other Common Ones! (And, of Course, How to Avoid Them)

"Too smart is dumb."

Old German Proverb

Key Concepts

- "Well packaged research is NOT truth!"
- There's no free lunch.
- Getting caught up in a fury of excitement.
- Falling for a scam.
- Options are risk-free.
- Trying to run first.
- Forgetting to negotiate commissions.
- Not looking over your shoulder.
- Managing your broker.

Here is an actual headline from a publication put out by a now defunct but what was at one time a very large and very profitable options brokerage firm:

"Soaring Sugar Prices Could Mean Stunning Profits for Speculators"

The subhead read: "A move of this magnitude could allow you to multiply your initial investment many times over."

How would you react if you receive a call from a very pleasant person who seems extremely interested in your financial well-being. In the course of the conversation, this person asks what your financial aspirations are. Your dreams? What you would do if you could add another $10,000 or $100,000 to your annual income? Here's what the conversation might sound like.

Salesperson: "Living in Coral Gables must be very pleasant with the beaches—the oceans and the endless sunshine. Sure beats winter's snow and slush. Do you take advantage of the ocean at your doorstep?"

Prospect: "Well, we do occasionally. The beaches are great. I'd like to do more fishing."

Salesperson: "What kind of boat do you have?"

Prospect: "We don't have one at the moment."

Salesperson: "If you could own any boat you wanted—if money was no object—what kind of boat would you buy?"

Prospect: "Gee—I don't know. There are so many great ones down here. I guess I'd go for one of those fishing yachts. You know what I mean—half pleasure boat, but rigged for deep-sea fishing."

Salesperson: "I know exactly what you mean. It would be my choice too. I think I'd like something around 30-foot. How about you?"

Prospect: "That's a good size, but a new one would be very expensive."

Salesperson: "Let's not worry about cost yet. What all would you want on it? How many bunks? What type of galley? How about radar and fish finders?"

Prospect: "It's sounding better and better. I'd want plush furnishings and a couple of big diesel engines—and, of course, lots of fishing gear."

Salesperson: "Now we're cooking. I saw a rig like that not too long ago in a magazine. The price tag was just under $100,000. If I could show you how to turn a couple of thousand into over a hundred thousand for that yacht—would you be interested?"

Prospect: "Of course, but how risky is it?"

Salesperson: "With the investment I'm talking about, the risk is very limited—BUT THE PROFIT IS UNLIMITED! Is that the type of investment you're looking for?"

Prospect: "Tell me more—"

It doesn't take any imagination to see where this sales conversation is headed. The salesperson probes the prospect to learn how much cash is available and then begins selling the story on the potential of sugar options. The salesperson closes this conversation by promising to overnight all the details on the sugar trade.

What is wrong with this scenario? Was the headline a complete fabrication? Did it have any basis in fact? What about the telephone sales presentation? Was it all just a fish story?

The headline and accompanying story were well researched and had bases in fact. Sugar can be an extremely volatile market. It often makes spectacular moves, as anyone who remembers the world-class move it made in 1974. That year started with some bullish fundamentals. The year-end (1973–74) stock-consumption ratio was low—20 percent. The sugar beet crop in Europe was faced with some adverse growing conditions. Fundamentals might account for 20- or 30-cent sugar, but it eventually more than doubled to over 65 cents. There was even talk of dollar-per-pound sugar. In the final months of this wild bull rally, futures prices shot up 36 cents a pound in 2 months. Thirty-six cents is three times higher than the prior post-World War II high! Then sugar came down faster than it went up, leveling in the 15- to 20-cent range. A similar market "panic" occurred in 1980. It was not as severe, but sugar rallied to 45 cents that time.

The sugar story called for 20-cent-per-pound sugar. At the time, sugar was trading around the 14-cent level. An increase to 20 cents would have been 6 cents times 112,000 pounds per contract, or $6,720, for an option that was in-the-money when the move began. The story did fairly state: "If the sugar market fails to move or moves against you, you could lose all or part of your purchase price. But your risk is strictly limited to the total purchase price of your options—a fraction of the underlying futures contract's actual market value." The fact that this company's commissions were large by industry standards and were included in "total purchase price" was not spelled out.

On the follow-up call, the salesperson ties it all together—the dream boat, the volatile sugar market, and outstanding fundamental reasons why a 6-cent move is well within expectations. Account papers were in the overnight packet with the sugar article. Also included was a return overnight envelope. The salesperson matched an appropriate number of options to the maximum cash available and had the prospect overnight everything back, including a cashiers check.

What did the prospect buy? In my opinion, nothing but a dream. The year being discussed was 1989 and sugar traded in the 14- to 15-cent range. All the options would have expired worthless.

Buying a dream—rather than working hard to make a well-balanced, diversified investment in options—is the biggest single mistake the majority of novice option traders make. All too often, options trading is presented as being a simple, limited risk investment.

Think about all that was said so far in this book about the law of probability, the distribution of winning and losing trades, the random nature of price activity, the need to survive losing trades to be in the right market, on the right side, at the right time. These are the realities of options trading! To think you can select a single trade and buy a dozen or so options that will net you $100,000 the first time you trade is not credible.

Options trading is a complex, highly speculative investment. Spectacular returns are possible, but they don't normally come easily. Professional traders, who make the above-average returns you have heard about, earn them.

The second biggest reason new traders lose, in my opinion, is they become overwhelmed by the research provided to them. Well-substantiated—and well-presented—projections based on historical patterns take on the aura of inevitability. Analysts take great pains drawing parallels to past moves they hope will be mirrored again. For example, the 1989 sugar situation was compared to those of 1974 and 1980. It seemed inconceivable that anything but a major bull move would occur.

Traders, even very experienced ones, get caught up in the excitement. Once the "facts" get circulated, everyone is talking about them. Herd psychology prevails. For a while, expectations often become self-fulfilling prophecies. Susceptibility to this phenomenon is another reason traders lose.

HOW VULNERABLE ARE YOU TO MASS HYSTERIA?

We often think of ourselves as being too logical or level-headed to be turned around by some bizarre social reality. We never bought a Hoola Hoop! We know Elvis is truly dead.

We all are vulnerable to some extent and we all have our Achilles' heels. If you understand how these phenomena of herd psychology work, you'll be better prepared to deal with them.

And as an option trader, you must be able to see reality for what it is in order to profitably trade reality. Let's start with a few examples to illustrate the point.

In London in 1524, the city awaited its doom. Numerous astrologers and fortune-tellers had foretold a massive flood. The great Thames, one of the world's best-behaved rivers, was to jump its bank and engulf the entire city on February 1. Doom and gloom was predicted by just about anybody anyone would listen to. Over 20,000 citizens, nearly the entire population, moved to higher ground.

On February 1, nothing happened. But the charlatans had an answer. A minor miscalculation had occurred. The flood was coming in 1624. Londoners were safe at least for awhile. Sound like any economic forecasters you know?

If you think these things only happened in the middle ages and that only the "poor unwashed masses" are caught up, think again. The "red bating" frenzy of McCarthyism even caught up the ACLU. That is a group who considers itself above emotionalism. The fact is it too was swept along in the 1950s and refused to defend suspected communists.

This type of behavior occurs in options trading regularly. Traders get caught in the power of mass delusion—"Prices are going through the ceiling!" "There's no end to this bull move!"—that causes them to lose control of their trading.

Let's take a quick look at how these fantasies develop. Our personal and collective body of beliefs can be roughly divided into two distinctly different types. The first is deeply rooted in physical fact. If we touch something very hot with our bare hand, we get burned. It is extremely difficult for anyone to get us to change our minds regarding these beliefs.

When we get into very complicated areas in which we have strong feelings, yet our understanding is somewhat vague, we tend to rely on others that we consider to be authorities. In the case of religion, it is ministers or priests. In politics, it is politicians. In finance, it is bankers, advisors, etc.

Options trading is no different. It certainly can be complex and vague. We look to professional traders, journalists, brokers, newsletter writers, etc. When they get caught up in the image of what "everyone" thinks is happening, they whip traders into a wild frenzy. This pulls the average investor in off the street. Everyone thinks the market will never stop. Novice option traders get burnt!

You need the discipline that comes with an understanding of the mass psychology that can overcome even the best traders' common sense. Add to this the courage to stand up to anyone trying to bully you or dazzle you into a trade. As noted earlier, successful traders control their emotions.

Another pitfall traders can encounter is old-fashioned fraud. This can involve classic Ponzi schemes, falsified research, or a creative presentation of research. For example, a dishonest broker locates 50 or so prospects who will listen to his pitch. He tells them he is backed by an excellent market analyst who has an incredible system for anticipating the direction of the S&P 500 Index. To demonstrate just how accurate it is, this pitchman offers to share the research with each of the 50 prospects for the next four weeks. With each weekly call, the prospects are told whether the system is calling for higher or lower prices. Half of the group are told higher and the other half lower.

Despite what happens in the S&P, at the end of four weeks there are at least six people who have seen four perfect market projections. These six are convinced and could be taken advantage of.

This is only one of the many, many scams perpetrated against would-be investors. There tends to be a common thread among these types of sales presentations, and that is an unbalanced presentation between the possibility of loss and the easy gains that can be made.

If you are solicited by a stock or futures broker who is using some type of promotion that delivers an unbalanced presentation of the enormous profits to be made from the market with little or no mention of risk, you'll know at least two important facts. First, the material is in violation of federal regulations. Second, beware of whoever is presenting the materials. If the initial overture is so out of tune, the rest of the song will probably be as well. The next refrain most likely will be a high-pressure sales presentation, consisting of overnight movement of account papers and checks. Consider yourself forewarned and keep your distance.

Advertising that is balanced and informative can be very helpful to you in deciding both if you are suited for options trading and if you want to learn more. As you absorb these types of messages, you'll learn how much risk you'll be accepting and have a more complete picture of profit opportunities.

You'll often see a big difference between the ethical broker and one who plays fast and loose. The former will spend time with you, send you information such as newsletters, and be sincerely interested in learning about your investment needs. The wheeler-dealer's objective is to raise the prospect's greed level to such a fever pitch that he or she buys "the dream," rather than reality. This is high-pressure sales. There's no time for thought and reflection. You'll hear lines like this:

- "An opportunity this good only comes by once in a blue moon!"
- "If you don't get $2,000 to me by noon tomorrow, you'll miss your chance to turn it into $10,000 in the next two weeks!"
- "Gold is going sky high—starting tomorrow!"

From ethical brokers, you'll hear descriptions of opportunities won and opportunities lost. The legitimate sales representative will tell you that sooner or later another good, solid trading opportunity will appear. You'll also be provided with a balanced presentation of the risk, something you won't hear from deal peddlers. Remember the old adage: "An option trade that sounds too good to be true probably is."

In its booklet *Before You Say Yes*, the NFA lists 15 questions that can help you distinguish qualified investment salespeople from swindlers. The following is a paraphrased version of those questions:

1. What is the commission rate? Other costs?
2. What are the risks?
3. Can you send me a written explanation of the risk involved?
4. Can you send me copies of your sales literature?

5. Are you selling an investment sold on a regulated exchange?
6. Which governmental or industry regulator supervises your firm?
7. Specifically, where would my money be held?
8. When and where can we meet in person?
9. How and when can I liquidate my investment if I so desire?
10. Will you send me your entire proposal in the mail?
11. If a dispute arises, what are the means available to resolve it?
12. Where did you get my name?
13. Would you mind explaining your proposal to my lawyer?
14. Would you give me the names of your principals and officers?
15. Can you provide references?

Experience has shown that the dishonest salespeople usually resist or are not prepared for this type of interrogation. Their "marks" are impulsive buyers who make entirely emotional decisions. The answers you get are vague and evasive. For example,

Question: How much are commissions?
Response: They're never a problem.
Question: If I have a complaint, what can I do?
Response: Don't even worry about that. I'll take care of you personally.

By gathering a lot of information and taking some time making your decision, you can protect yourself from the majority of swindlers—but not all. The very best mimic legitimate sales operations, but most can be uncovered if you do your homework and control your emotions.

Another simple procedure you can do to avoid scam artists is call the NASD or the NFA Information Center and ask these questions:

1. Is the person you've talked with registered?
2. With which firm is the person registered?
3. How long has he or she been registered?

4. Has the NASD or NFA taken any public disciplinary action against the person?

5. Has he or she been registered with any firm that has been cited for compliance problems?

6. Is there anything else you can tell me about the person, since I'm considering investing with him or her?

Once you decide who you think the right broker is, you make your selection by completing the account papers. Your broker should ask you to read them carefully and fill them out in your own handwriting. The reason for you to complete it in your own handwriting is so the broker is assured that you saw them and had the opportunity to read them. The broker will ask you if you understand them and have any questions. The account papers describe the risks you are assuming. You have no one to blame but yourself if you ignore them or sign them without reading them carefully.

If the broker offers to fill them out for you ("I'll just get them typed up for you"), consider this a warning signal. The easier the broker makes it for you to fill out the account papers, the less likely you are to study them. They are worth the time and effort it takes to understand what you are signing because they define the risk you face in great detail.

Poor communications is another miscue that often occurs between clients and their brokers. It's important that you and your broker set up some guidelines. Here are some to use to sidestep this problem. Be sure to cover them all before trading begins.

1. Time of day you are most easily accessible.

2. Whether you are to call in or whether your broker will call you.

3. What steps will be taken if you can't be contacted; i.e., should your broker put stops in the market, close out positions after 4, 8, 12, or 24 hours of not being in contact, etc.?

4. What steps will be taken if the account goes into debit? Do you plan to wire money into the account? Transfer from another account? Overnight a check?

5. How trading emergencies will be handled.

6. What hours you are available and whom to contact when you are not available.

7. Normal office hours, plus any evening hours your broker is available.

8. You may need to provide your broker with your home or office phone number, if that's the only way for your broker to stay in touch with you. Also, be sure to ask for hers or his.

9. Your broker needs to know whom to contact if you are not always near a phone. This person should be able to reach you quickly. For example, farm wives are often in radio contact with their husbands during planting and harvesting. Or many businesspeople carry cellular phones when out of their offices.

10. You need to make provisions for unusual situations, such as when you are traveling or taking a vacation.

The point is simply this: It is easy to avoid the mistake of losing contact with your broker. If a situation arises in which you are going to be out of touch, either put stops in the market, close out the positions, or even close your account. Do not risk trading when you cannot be reached quickly, if the market situation warrants it. If your broker doesn't mention this, be sure you do.

Another common misfire made by new option traders is jumping the gun. Many traders come to the conclusion there is no need to hesitate since the risk of buying a put or a call can be predetermined. They reason: "The option only costs $900, or $1,000 including fees and commissions. I can afford that. Go ahead and buy it."

First, if the option expires worthless, you lose 100 percent of your investment. Think of it in those terms, rather than a dollar amount you can stand to lose.

When you think of it as 100 percent, you start trying to figure how can you salvage some of it. What about cutting the loss to 50 percent? Or 20 percent? It's this kind of thinking that leads to the use of trailing stops behind your positions.

The opposite—waiting too long to trade—can be equally debilitating. Some people wait until they learn everything there is to know about options trading before they place their first trade. Of course, this never happens.

On the other hand, if you have a gut feeling that options may not be for you, don't fight it. You don't have to trade. Trading stock and futures options is highly speculative, not suited for every investor. Never let a broker or "friend" bully or dare you into trading. Hang up on any broker who challenges like this:

"If you don't have the guts for this trade, you better get a testosterone check!"

"If you have to ask your husband for permission to trade, you're not a '90s woman!"

Your motivation for placing an options trade should be profit based on a sound, unemotional evaluation of the strategy, the market, and the trade's risk-to-reward ratio.

A variation of these last scenarios is the purchasing of an option at the wrong strike price or month because you don't have enough money in your account to buy the right one. For example, let's say you are considering buying a December crude oil call. The futures price is $20.54 per barrel and you're looking at the $20.00 strike price. It's at $0.74 or $740 (1,000 bbl × $0.74). Since you only have $500 in trading equity, you settle on a $21.00 per barrel strike price. It's at $0.16 or $160 plus commissions and fees.

This is a big change, from approximately 50 cents in-the-money to 50 cents out. The in-the-money call had $500 of intrinsic value and the rest was for time to expiration. Since the out-of-the-money option has less time value than the in ($160 versus $240), the market is telling you time is not valuable. But the key error is constructing your strategy to accommodate your wallet. Also, you must resist the suggestion of your broker to downtrade in these circumstances. Brokers normally work on commissions and are continuously looking for ways to sell a trade.

You're usually better off searching for a new trade that meets all your criteria and also is affordable—or standing aside the market. Standing aside is a totally acceptable stance. Inexperienced traders often make the mistake of overtrading, feeling they should always have something going in order to be successful. The pros often sit on the bench when the game is not going their way or when they are not sure which way it is headed. This leads to the next subject, which is the proper way to get into the markets in the first place.

CHAPTER 12

Getting Started, or the Secrets of Successful Options Trading

"What we call the beginning is often the end. And to make an end is to make a beginning. The end is where we start from."

T. S. Eliot
(Little Gidding, 1942)

Key Concepts

- The Secret behind the Secret
- Master Trader, Journeyman, Loser—Which Will You Be?
- Option Traders Checklist
- Getting Started the Right Way

What is the secret to successful trading? Just the word "secret" should tell you something. If there is a secret to trading and anyone told it, it wouldn't be a secret—would it?

There is also the problem of greed. All option traders have it to one degree or another. It helps us overcome the fear of loss, allowing us to enter the markets. Since we do have a healthy dose of it, would any successful trader share a secret that could open the door to untold wealth? For free? Especially if sharing the secret could very well reduce or destroy the effectiveness of it? Keep these thoughts in mind when you read books or articles offering trading "secrets."

The secret to options trading is the same as the secret of success in any field. There are two basic aspects—one involves nature, the other nurture. For good measure, you can throw in some good old-fashioned luck.

WHAT DO A SUCCESSFUL FASHION DESIGNER AND AN NFL LINEBACKER HAVE IN COMMON?

First of all, you'll say they are very good at what they do. The designer attended some outstanding schools, worked for years under a well-known "rabbi," and got a big "break"—catapulting him or her to the top. The linebacker sweated out 10 to 12 seasons on a variety of levels to become an NFL "rookie." After two or three years in the pros without a career-ending injury (luck), he's at the Pro or Super Bowl.

In both cases, years were spent in study, learning the rules. But it took more than that. To become superstars, they had to have a special knack for knowing what excites the fashion world or which way a misdirection play is really headed. Without this special ability, this sixth sense, these two individuals could have very nice careers in their chosen fields, but on more modest levels. Also, without some luck, they may have drifted to other fields in which they may not be as well suited.

This analysis additionally needs to take into consideration accidents of nature. If the potential designer was born color-blind, how would this diminish chances for success? Or the football player, what if he had a club foot? They might have the hearts of lions, but certain physical and mental faculties are usually necessary for them to compete successfully. For the option trader, math skills and a good memory are a must. If you can't do simple math quickly and accurately in your head, you'll be at a distinct disadvantage, since you're constantly having to figure the value of a spread or converting a 40-point move in sugar to dollars and cents ($11.20 = 1¢; $11.20 × 40¢ = $448).

To make all the spur-of-the-moment mathematical calculations, you need to be able to immediately recall hundreds of facts and figures. Additionally, I have met few successful traders who were not able to recollect past price patterns and moves. History repeats itself daily in the pits. Your mind must become an encyclopedia of trivia about the markets you trade. This is another reason I often caution new traders to concentrate on only a few stock issues or commodity markets in the beginning.

The secret of success also entails performing the hard work needed to learn the rules others have found necessary. I shared many of these with you in an earlier chapter, but no one can teach you the discipline you need to follow them. This is part of the nurturing aspect of learning to trade.

Let's jump to the natural or intuitive part of trading. How do you know if you have it? What's the test? If you don't have "it," is there any hope?

As with any profession, there at least two levels of proficiency—the masters and the journeymen. In this industry, the masters are called market wizards. You should read Jack Schwager's two books on market wizards for inspiration and information. They contain a series of interviews with a group of recognized trading masters.

If you're not a master, do you want to be a journeyman? When you seriously consider it, it is not all that bad. A journeyman option trader follows the rules, trades when the risk-to-reward ratio is well in his or her favor, and primarily stays with the more conservative strategies (spreads, for example). The journeyman's profit objective, by option trading standards, is also modest—usually in the 25 to 50 percent return to equity per annum. By other investment standards, it isn't modest.

Can any trader become a journeyman?

Unfortunately, no! There is one other class of individuals also present in all other vocations or avocations. I am referring to the losers.

How do you know if you're likely to be a loser in options trading? No one, of course, knows for sure, but there are some telltale characteristics. Even before they trade, the losers often know in their hearts that they are going to lose. A negative personality rarely earns profits consistently. Losers are usually attracted to options for the wrong reasons—to make a lot of money fast without exerting much effort. Therefore, they don't spend the time required to learn some of the more complicated strategies that are more conservative by comparison.

This lack of patience also is reflected in the inability to wait for a trade with an above-average risk-to-reward ratio. Losers are impatient and act very impulsively. They usually refuse to keep trading journals and rarely learn from their mistakes. Brokers can strongly influence them by offering quick fixes to lagging winnings, which usually compounds the problem. Loss of independence in the options game often equates to loss of capital.

Success requires focus, patience, and the ability to act fast and decisively without losing control of your emotions, trading too many markets, or refusing to wait for strategies to succeed. Quite frankly, options trading can sometimes be tiresome. When you buy an option, you buy time—time for the market to move as you anticipate. And that can take awhile. For example, trades may stay open 2 or 3 months. During the first 30 to 60 days, there may be an insignificant change in your position. You must be prepared and patient. While waiting for something to happen, you can lose faith, in both yourself and your trades. Persistence of purpose is another virtue of successful people.

One clue that you have the potential to be a master, or at least an upper-level journeyman, is your gut feeling. As you read this book, do you hear yourself saying: "I can do that!" Or do you keep hearing the refrain—"This is too risky. This is too risky. Do I really want to tackle something this speculative?"

There is nothing wrong with either reaction. If in your soul you're not comfortable with all you've learned so far—give it up! Thank your lucky stars you didn't pay $500, $1,000, or $10,000 for the same lesson, as others have. Take a philosophic approach to the experience. You bought a book, you learned a lot about a unique type of investment, and you got your money's worth.

On the other hand, if you're chomping at the bit to fill out account papers and attack the markets at this stage, take a cold shower. Then write down the answers to these questions:

1. Why are you psychologically and financially suited to trade options?
2. What are your goals? Are they realistic? Specific?
3. How do you want and plan to use options to reach your objectives?
4. What are your limitations, and how do you plan to manage them?
5. Will you trade stocks or futures options?

6. Which stock or futures markets do you plan to trade? Describe your prospective portfolio.

7. How will you analyze the markets? Will you trade for the long, medium, or short term?

8. What system will you use to select trades, entry-exit price points, trailing protective stops, trading volume, number (size) of positions, type of orders, software, price quotation source or equipment, etc.?

9. Do you have a clear understanding of money management? Do you have a written set of rules? Do you fully understand the concept of survivability as it pertains to options trading?

10. What do you need or want from a broker and his or her firm?

11. Are you comfortable with all the nitty-gritty that goes along with options trading, like understanding the account papers, your rights and responsibilities, daily and monthly statements, etc.?

Once you have answers to these questions—again, I strongly recommend putting them in writing because it clarifies thinking— you're ready to begin the broker selection process. (See Figure 12–1.)

As you are searching for a good broker, I'd recommend one additional exercise before you fund an actual trading account. Consider doing some practice or paper trading.

There are some computer programs that simulate options trading you might want to check into. One is available from COMEX at a cost of $69. It's called The Game. You start with $10,000 in equity and trade COMEX gold and silver options.

Before you invest in The Game, you might want to call the Coffee, Sugar and Cocoa Exchange for a free copy of its program called SOFTSware. It's divided into three parts. The first part, "About Options," provides a brief overview of options terminology and strategy. It includes the contract specifications for coffee, sugar, and cocoa options and contains sources for more information.

The second and third sections will be the ones most useful to you. For example, "Choose a Strategy," which is the second part, prompts you to pick a commodity to trade. Naturally, your choices

F I G U R E   12–1

The Trading Process

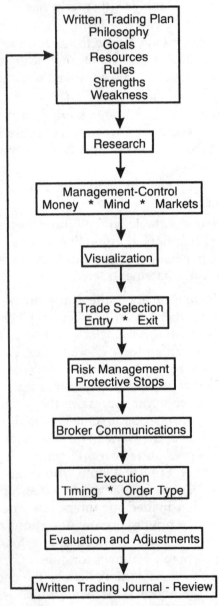

Trading is called a process because it is on-going. It never stops, just as the markets are continuously in flux.

are coffee, sugar, or cocoa. Next you select a market opinion: bullish, bearish, or neutral. At the next screen, you pick a strategy appropriate to that type of market. You are presented with several different strategies to test. If your market opinion choice was neutral, for example, you could try any one of the following strategies:

- "A" Butterfly
- "V" Butterfly
- "A" Condor
- "V" Condor
- Long Straddle
- Short Straddle
- Long Strangle
- Short Strangle
- Conversion
- Box
- Call Ratio Write

You have similarly appropriate selections for bullish or bearish market sentiments.

Once a strategy is chosen, the next screen provides an explanation and example of the trade. You then have four more information screens. One gives you details of the profit and loss of the trade, at expiration, based on a wide variety of prices for the underlying commodity. The next screen graphs the trade. This is a valuable picture of the profit and loss potential for each of the many strategies illustrated. A time-volatility screen lets you change the time to expiration, volatility rate, and/or interest rate. You can see the impact of changes in volatility in dollars and cents. If you're not pleased with your strategy selection, you can switch to a new one and compare profitability.

The last part of SOFTSware gives you the opportunity to build your own strategy. You can combine up to four different options into a strategy after selecting the commodity and market direction. You fill in the following blanks:

1. Long or short
2. Call, put, or futures
3. Strike or futures price

4. Number of contracts

5. Option premium

The program then calculates the four screens previously described for the "Choose Strategy" section. You can then make refinements (change, volatility, etc.) at will. This portion is a great way to test yourself to make sure you understand how and when to use a bull spread, condor, butterfly, etc.

This program is designed to promote the Coffee, Sugar and Cocoa Exchange. These are exciting markets to trade, but sometimes liquidity is not as high as one would wish for when trading the first time. Always be cautious of thin markets, no matter what exchange.

WHAT ELSE SHOULD YOU DO TO PREPARE, BESIDES TAKING ANOTHER COLD SHOWER?

Consider investing in a seminar or two. Most of the exchanges, stock and futures, offer courses on a variety of educational levels. Contact those nearby or the ones you think you'll utilize in your trading. If you plan to trade options-on-futures, contact the Chicago Merc or the CBOT. If you're into stock options, contact the appropriate stock exchange nearest to you. (See Appendix 3 for addresses and telephone numbers.)

Two final admonitions: First, there is only one way to really understand options trading, and that is by doing it. You can spend the rest of your days studying it, but until you place your first trade, you'll never experience what all the excitement is about.

Second, and last of all, never forget the simple fact that you, and everyone else trading options, are attempting to anticipate what will happen 5 minutes, 5 weeks, or 5 months from the time you place a trade. No one knows the future. Therefore, you must never marry a position and you should take what the market gives you.

Glossary of Terms

Actuals The physical or cash commodity, as distinguished from commodity futures contracts.

Administrative Law Judge (ALJ) A CFTC official authorized to conduct a proceeding and render a decision in a formal complaint procedure.

Aggregation The policy under which all futures positions owned or controlled by one trader or a group of traders are combined to determine reporting status and speculative limit compliance.

All or None (AON) An order which must be filled in its entirety or not at all.

American-Type Option An option which may be exercised at any time prior to expiration.

Arbitrage The simultaneous purchase of one commodity against the sale of another in order to profit from distortions from usual price relationships. See also *Spread*; *Straddle*.

Arbitration A forum for the fair and impartial settlement of disputes that the parties involved are unable to resolve between themselves. NFA's arbitration program provides a forum for resolving futures-related disputes. NASD handles stock-related complaints.

Assignment Notice to an option writer that an option has been exercised by the option holder. This can happen at any time during the life of an option with American-type options and only at or near expiration for European-style options.

Associated Person (AP) An individual who solicits orders, customers, or customer funds on behalf of a futures commission merchant, an introducing broker, a commodity trading advisor, or a commodity pool operator and who is registered with the Commodity Futures Trading Commission (CFTC) or the National Futures Association (NFA).

At-the-Money An option whose strike price is equal to the market value of the underlying futures contract. Can also refer to an order to buy a futures contract at the current bid-ask price. See *Market Order*.

Automatic Exercise The exercise by the clearinghouse of an in-the-money option at expiration, unless the holder of the option submits specific instructions to the contrary.

Award See *Reparations Award*.

Back Spread A spread in which more options are purchased than sold, and where all options have the same underlying entity and expire at the same time. Back spreads are usually delta-neutral.

Backwardation A futures market where the more distant delivery months for a commodity trade at a discount to the near-term delivery months.

Basis   The difference between the cash or spot price and the price of the nearby futures contract.

Bear Market (bear/bearish)   A market in which prices are declining; a market participant who believes prices will move lower is called a "bear." A news item is considered "bearish" if it is expected to produce lower prices.

Bear Spread   Any spread in which a decline in the price of the underlying entity will increase the value of the spread.

Beta   The measure of how the options market correlates to the movement of the underlying futures market.

Bid   An offer to buy a specific quantity of a commodity at a stated price.

Board of Trade   Any exchange or association of persons who are engaged in the business of buying or selling any commodity or receiving the same for sale on consignment. Usually means an exchange where commodity futures and/or options are traded. See also *Contract Market*; *Exchange*.

Box   A long call and short put at one exercise price, and a short call and a long put at a different exercise price. All four options must have the same underlying entity and expire at the same time.

Break   A rapid and sharp price decline.

Broker   A person who is paid a fee or commission for acting as an agent in making contracts or sales; a floor broker in commodities futures trading; a person who actually executes orders on the trading floor of an exchange; an account executive (associated person), the person who deals with customers and their orders in commission house offices. See *Registered Commodity Representative*; *Registered Representative* (stockbroker).

Brokerage   A fee charged by a broker for execution of a transaction, an amount per transaction, or a percentage of the total value of the transaction; usually referred to as a commission fee.

Bucket, Bucketing   The illegal practice of accepting orders to buy or sell without executing such orders; and the illegal use of the customer's margin deposit without disclosing the fact of such use.

Bull Market (bull/bullish)   A market in which prices are rising. A participant in futures who believes prices will move higher is called a "bull." A news item is considered "bullish" if it is expected to bring on higher prices.

Bull Spread   Any spread in which a rise in the price of the underlying entity will theoretically increase the value of the spread.

Butterfly   The sale (purchase) of two identical options, together with the purchase (sale) of one option with an immediately higher exercise price and one option with an immediately lower exercise price. All options must be of the same type, for the same underlying entity, and expire at the same time.

Buy or Sell on Close or Opening   To buy or sell at the end or the beginning of the trading session.

Buy/Write   The purchase of an underlying contract and the simultaneous sale of a call.

Buyer   The purchaser of an option, either a call option or a put option. Also referred to as the option holder. An option purchase may be in connection with either an opening or a closing transaction.

Buying Hedge (or Long Hedge)   Buying futures contracts to protect against the possible increased cost of commodities which will be needed in the future. See *Hedging*.

Calendar Spread   See *Time Spread*.

Call (Option)   The buyer of a call option acquires the right but not the obligation to purchase a particular futures contract at a stated price on or before a particular date. Buyers of call options generally hope to profit from an increase in the futures price of the underlying commodity.

Car(s)   This is a colloquialism for futures contract(s). It came into common use when a railroad car or hopper of corn, wheat, etc., equaled the amount of a commodity in a futures contract. See *Contract*.

Carrying Broker   A member of a commodity exchange, usually a clearinghouse member, through whom another broker or customer chooses to clear all or some trades.

Carrying Charges   Costs incurred in warehousing the physical commodity, generally including interest, insurance, and storage.

Carryover   That part of the current supply of a commodity consisting of stocks from previous production/marketing seasons.

Cash Commodity   Actual stocks of a commodity, as distinguished from futures contracts; goods available for immediate delivery or delivery within a specified period following sale; or a commodity bought or sold with an agreement for delivery at a specified future date. See *Actuals*; *Forward Contracting*.

Cash Forward Sale   See *Forward Contracting*.

Certificated Stock   Stocks of a commodity that have been inspected and found to be of a quality deliverable against futures contracts, stored at the delivery points designated as regular or acceptable for delivery by the commodity exchange.

Charting   The use of graphs and charts in the technical analysis of futures markets to plot trends of price movements, average movements of price volume, and open interest. See *Technical Analysis*.

Christmas Tree   A type of ratio vertical spread where options are sold at two or more different exercise prices.

Churning   Excessive trading of the customer's account by a broker, who has control over the trading decisions for that account, to make more commissions while disregarding the best interests of the customer.

Class of Options   All call options—or all put options—on the same underlying stock or futures contracts.

Clearing   The procedure through which trades are checked for accuracy, after which the clearinghouse or association becomes the buyer to each seller of a futures contract and the seller to each buyer.

Clearing Member   A member of the clearinghouse or association. All trades of a non-clearing member must be registered and eventually settled through a clearing member.

Clearing Price   See *Settlement Price.*

Clearinghouse or Clearing Corporation   An agency connected with exchanges through which all futures and option contracts are made, offset, or fulfilled through delivery of the actual commodity and through which financial settlement is made; often, a fully chartered separate corporation rather than a division of the exchange proper. Once a trade has been cleared, the clearing corporation becomes the buyer to every seller and the seller to every buyer.

Close (the)   The period at the end of the trading session, officially designated by the exchange, during which all transactions are considered made "at the close."

Closing Purchase Transaction   An exchange transaction in which the purchaser is reducing or eliminating the short position in the option series involved in the transaction.

Closing Range   A range of closely related prices at which transactions took place at the closing of the market; buy and sell orders at the closing might have been filled at any point within such a range.

Closing Sale Transaction   An exchange transaction in which the seller is reducing or eliminating the long position in the option series involved in the transaction.

Combination   A trading strategy. Combinations involve buying or selling both a put and a call on the same stock or futures. A position created either by purchasing both a put and a call or by writing both a put and a call on the same underlying entity.

Commission   (1) A fee charged by a broker to a customer for performance of a specific duty, such as the buying or selling of futures contracts or stock issued. (2) Sometimes used to refer to the Commodity Futures Trading Commission (CFTC).

Commodity   An entity of trade or commerce, services, or rights in which contracts for future delivery may be traded. Some of the contracts currently traded are wheat, corn, cotton, livestock, copper, gold, silver, oil, propane, plywood, currencies, and Treasury bills, bonds, and notes.

Commodity Exchange Act   The federal act that provides for federal regulation of futures trading.

Commodity Futures Trading Commission (CFTC)   A commission set up by Congress to administer the Commodity Exchange Act, which regulates trading on commodity exchanges.

Commodity Pool   An enterprise in which funds contributed by a number of persons are combined for the purpose of trading futures contracts and/or options on futures. Not the same as a joint account.

Commodity Pool Operator (CPO)   An individual or organization that operates or solicits funds for a commodity pool. Generally required to be registered with the Commodity Futures Trading Commission.

Commodity Trading Advisor (CTA)   An individual or firm that, for a fee, issues analyses or reports concerning commodities and advises others about trading in commodity futures, options, or leverage contracts.

Complainant   The individual who files a complaint against another individual or firm.

Condor   The sale (purchase) of two options with consecutive exercise prices, together with the purchase (sale) of one option with an immediately lower exercise price and one option with an immediately higher exercise price. All options must be of the same type, have the same underlying entity, and expire at the same time.

Confirmation Statement   A statement sent by a commission house or broker-dealer to a customer when a futures, options, or stock position has been initiated or some other transaction has taken place. The statement shows the number of contracts (stock shares) bought or sold and the prices at which the contracts were bought or sold. It might also show any funds moved in or out of the account. Sometimes combined with a purchase and sale statement.

Consolidation   A pause in trading activity in which price moves sideways, setting the stage for the next move. Traders evaluate their positions during periods of consolidation.

Contango   A futures market where the more distant delivery months for a commodity trade are at a premium to the near-term delivery months.

Contingency Order   An order which becomes effective only upon the fulfillment of some condition in the marketplace.

Contract   A term of reference describing a unit of trading for a commodity.

Contract Grades   Standards or grades of commodities listed in the rules of the exchanges which must be met when delivering cash commodities against futures contracts. Grades are often accompanied by a schedule of discounts and premiums allowable for delivery of commodities of lesser or greater quality than the contract grade.

Contract Market   A board of trade designated by the Commodity Futures Trading Commission to trade futures or option contracts on a particular commodity. Commonly used to mean any exchange on which futures are traded. See also *Board of Trade*; *Exchange*.

Contract Month   The month in which delivery is to be made in accordance with a futures or options contract.

Controlled Account   See *Discretionary Account*.

Corner   To secure control of a commodity so that its price can be manipulated.

Correction   A price reaction against the prevailing trend of the market. Common corrections often amount to 33, 50, or 66 percent of the most recent trend movement. Sometimes referred to as a "retracement."

Cost of Recovery   Administrative costs or expenses incurred in obtaining money due the complainant. Included are such costs as administrative fees, hearing room fees, charge for clerical services, travel expenses to attend the hearing, attorney's fees, filing costs, etc.

Cover   To offset a previous futures transaction with an equal and opposite transaction. Short covering is a purchase of futures contracts to cover an earlier sale of an equal number of the same delivery month; liquidation is the sale of futures contracts to offset the obligation to take delivery on an equal number of futures contracts of the same delivery month purchased earlier.

Covered Option   An option that is written against an opposite position in the underlying stock, futures contract, or commodity at the time of execution or placement of the order.

Covered Writer   The seller of a covered option, put or call.

Credit   Money received from the sale of options.

Current Delivery (Month)   The futures or options contract which will come to maturity and become deliverable during the current month; also called "spot month."

Customer Account   An account established by the clearing member solely for the purpose of clearing exchange transactions by the clearing member on behalf of its customers other than those transactions of a floor trader.

Customer Segregated Funds   See *Segregated Account.*

Day Order   An order that, if not executed, expires automatically at the end of the trading session on the day it was entered.

Day Traders   Traders who take positions and then liquidate them prior to the close of the trading day.

Dealer Option   A put or call on a physical commodity, not originating on or subject to the rules of an exchange, written by a firm that deals in the underlying cash commodity.

Debit   Money paid for the purchase of options.

Debit Balance   Accounting condition where the trading losses in a customer's account exceed the amount of equity in the customer's account.

Deck   All of the unexecuted orders in a floor broker's possession.

Deep-out-of-the-Money Options   Definitions vary by exchange. These are options with strike prices that are not close to the strike price nearest the current price of the underlying entity. A typical definition would be 2 strike prices, plus the number of calendar months remaining until the option expires, away from the strike price closest to the current value of the underlying stock issue or futures contract.

Default   (1) In the futures market, the failure to perform on a futures contract as required by exchange rules, such as a failure to meet a margin call or to make or take delivery. (2) In reference to the federal farm loan program, the decision on the part of a producer of commodities not to repay the government loan, but instead to surrender his or her crops. This usually floods the market driving prices lower.

Deferred Delivery   The distant delivery months in which futures trading is taking place, as distinguished from the nearby futures delivery month.

Deliverable Grades   See *Contract Grades.*

Delivery   The tender and receipt of an actual commodity or warehouse receipt or other negotiable instrument covering such commodity, in settlement of a futures contract.

Delivery Month   A calendar month during which a futures contract matures and becomes deliverable. Options are also assigned to delivery months.

Delivery Notice   Notice from the clearinghouse of a seller's intention to deliver the physical commodity against a short futures position; precedes and is distinct from the warehouse receipt or shipping certificate, which is the instrument of transfer of ownership.

Delivery Points   Those locations designated by commodity exchanges at which stocks of a commodity represented by a futures contract may be delivered in fulfillment of the contract.

Delivery Price   The official settlement price of the trading session during which the buyer of futures contracts receives through the clearinghouse a notice of the seller's intention to deliver, and the price at which the buyer must pay for the commodities represented by the futures contract.

Delta   The sensitivity of an option's theoretical value to a change in the price of the underlying entity.

Delta-Neutral Spread   A spread where the total delta position on the long side and the total delta position on the short side add up to approximately zero.

Diagonal Spread   A two-sided spread consisting of options at different exercise prices and with different expiration dates. All options must be of the same type and have the same underlying entity.

Discount   (1) A downward adjustment in price allowed for delivery of a quantity of a commodity of lesser than deliverable grade against a futures contract. (2) Sometimes used to refer to the price difference between futures or options of different delivery months, as in the phrase "July at a discount to May," indicating that the price of the July is lower than that of the May.

Discovery   The process which allows one party to obtain information and documents relating to a dispute from the other party(ies) in the dispute.

Discretionary Account   An arrangement by which the holder of the account gives written power of attorney to another, often a broker, to make buying and selling decisions without notification to the holder; often referred to as a managed account or controlled account.

Elasticity   A characteristic of commodities which describes the interaction of the supply of, demand for, and price of a commodity. A commodity is said to be elastic in demand when a price change creates an increase or decrease in consumption. The supply of a commodity is said to be elastic when a change in price creates a change in the production of the commodity. Inelasticity of supply or demand exists when either supply or demand is relatively unresponsive to changes in price.

Equity   The dollar value of an account.

European Option   An option which may only be exercised on the expiration date.

Exchange   An association of persons engaged in the business of buying and selling commodity futures, stocks, and options. See also *Board of Trade; Contract Market.*

Exercise   To exercise an option means the holder elects to accept the underlying stock or futures contract at the option's strike price.

Exercise Price   The price at which the buyer of a call (put) option may choose to exercise his or her right to purchase (sell) the underlying stock or futures contract. Also called strike price.

Expiration Date   Generally the last date on which an option may be exercised.

Extrinsic Value   The price of an option less its intrinsic value. The entire premium of an out-of-the-money option consists of extrinsic value. Also referred to as time value.

F.O.B.   Free on board; indicates that all delivery, inspection, and elevation or loading costs involved in putting commodities on board a carrier have been paid.

Fair Value   See *Theoretical Value.*

Feed Ratios   The variable relationships of the cost of feeding animals to market-weight sales prices, expressed in ratios, such as the hog-corn ratio. These serve as indicators of the profit return or lack of it in feeding animals to market weight.

Fence   A long (short) underlying position, together with a long (short) out-of-the-money put and a short (long) out-of-the-money call. All options must expire at the same time. When the fence includes a long (short) underlying position, it is sometimes known as a risk conversion (reversal).

Fibonacci Number or Sequence of Numbers   The sequence of numbers (1,1,2,3,5,8,13,21,34,55,89,144,233,–) discovered by the Italian mathematician Leonardo Fibonacci in the thirteenth century and the mathematical basis of the Elliott wave theory, where the first two terms of the sequence are 1 and 1 and each successive number is the sum of the previous two numbers.

Fiduciary Duty   The responsibility imposed by the operation of law (from congressional policies underlying the Commodity Exchange Act or the Security Act) which requires that the broker act with special care in the handling of a customer's account.

Fill Or Kill (FOK)   An order which must be filled immediately, and in its entirety. Failing this, the order will be canceled.

First Notice Day   First day on which notices of intention to deliver cash commodities against futures contracts can be presented by sellers and received by buyers through the exchange clearinghouse.

Floor Brokers   Individuals who execute orders on the trading floor of an exchange for any other person.

Floor Traders   Members of an exchange who are personally present, on the trading floor of an exchange, to make trades for themselves and their customers. Sometimes called scalpers or locals.

Forward Contracting   A cash transaction common in many industries, including commodities, in which the buyer and seller agree upon delivery of a specified quality and quantity of goods at a specified future date. Specific price may be agreed upon in advance, or there may be agreements that the price will be determined at the time of delivery on the basis of either the prevailing local cash price or a futures price.

Free Supply   Stocks of a commodity which are available from commercial sale, as distinguished from government-owned or controlled stocks.

Free Trade   An option spread initiated by purchasing a close to in-the-money put or call and later completed by selling a further out-of-the-money put or call of the same expiration period at the same premium. When completed, it requires no margin or equity.

Front Spread   See *Ratio Vertical Spread.*

Fully Disclosed   An account carried by the futures commission merchant in the name of the individual customer; opposite of an omnibus account.

Fundamental Analysis   An approach to analysis of price trends which examines the underlying factors that will affect the supply-and-demand equation and thus the price of a stock or futures contract. See also *Technical Analysis.*

Futures Commission Merchant (FCM)   An individual or organization that solicits or accepts orders to buy or sell futures contracts or commodity options and accepts money or other assets from customers in connection with such orders. Must be registered with the Commodity Futures Trading Commission or the National Futures Association.

Futures Contract   A standardized binding agreement to buy or sell a specified quantity or grade of a commodity at a later date, i.e., during a specified month. Futures contracts are freely transferable and can be traded only by public auction on designated exchanges.

Futures Industry Association (FIA)   The national trade association for the futures industry.

Futures-Type Settlement   A kind of settlement in which accounts are settled daily, mark-to-the-market. Excess cash can be withdrawn by owner.

Gamma   The sensitivity of an option's delta to a change in the price of the underlying entity.

Gap   A trading day during which the daily price range is completely above or below the previous day's range causing a gap between them to be formed. Some traders then look for a retracement to "fill the gap."

Good til Canceled (GTC)   An order to be held by a broker until it can be filled or canceled.

Grantor   A person who sells an option and assumes the obligation but not the right to sell (in the case of a call) or buy (in the case of a put) the underlying stock, futures contract, or commodity at the exercise price. Also referred to as writer.

Gross Processing Margin (GPM)   Refers to the difference between the cost of soybeans and the combined sales income of the soybean oil and meal which results from processing soybeans.

Guided Account   An account that is part of a program directed by a commodity trading advisor (CTA) or futures commission merchant (FCM). The CTA or FCM plans the trading strategies. The customer is advised to enter and/or liquidate specific trading positions. However, approval to enter the order must be given by the customer. These programs usually require a minimum initial investment and may include a trading strategy that will utilize only a part of the investment at any given time.

Hedging   The sale of futures contracts or options in anticipation of future sales of cash commodities as a protection against possible price declines, or the purchase of futures contracts or options in anticipation of future purchases of cash commoditiy or stock as a protection against increasing costs. See also *Buying Hedge*; *Selling Hedge*.

Holder   See *Buyer*.

Horizontal Spread   See *Time Spread*.

In-the-Money   An option having intrinsic value. A call is in-the-money if its strike price is below the current price of the underlying stock or futures contract. A put is in-the-money if its strike price is above the current price of the underlying stock or futures contract.

Inelasticity   A characteristic that describes the interdependence of the supply of, demand for, and price of an entity. A commodity is inelastic when a price change does not create an increase or decrease in consumption; inelasticity exists when supply and demand are relatively unresponsive to changes in price. See also *Elasticity*.

Initial Margin   A customer's funds required at the time a futures position is established, or an option is sold, to assure performance of the customer's obligations. Margin in commodities is not a down payment, as it is in securities. See also *Margin*.

Intermarket Spread   A spread consisting of opposing positions in instruments with two different underlying markets.

Intrinsic Value   The absolute value of the in-the-money amount, that is, the amount that would be realized if an in-the-money option were exercised.

Introducing Broker (IB)   A firm or individual that solicits and accepts commodity futures orders from customers but does not accept money, securities, or property from them. An IB must be registered with the National Futures Association and must carry all of its accounts through an FCM on a fully disclosed basis.

Inverted Market   A market in which the nearer months are selling at premiums over the more distant months; characteristically, a market in which supplies are currently in shortage.

Invisible Supply   Uncounted stocks of a commodity in the hands of wholesalers, manufacturers, and producers which cannot be identified accurately; stocks outside commercial channels but theoretically available to the market.

Kappa   See *Vega*.

Last Trading Day   The day on which trading ceases for the maturing (current) delivery month.

Leg   One side of a spread position.

Leverage   Essentially allows an investor to establish a position in the marketplace by depositing funds that are less than the value of the contract.

Leverage Contract   A standardized agreement calling for the delivery of a commodity with payments against the total cost spread out over a period of time. Principal characteristics include standard units and quality of a commodity and of terms and conditions of the contract; payment and maintenance of margin; close-out by offset or delivery (after payment in full); and no right to or interest in a specific lot of the commodity. Leverage contracts are not traded on exchanges.

Leverage Transaction Merchant (LTM)   The firm or individual through whom leverage contracts are entered. LTMs must be registered with the Commodity Futures Trading Commission.

Life of Contract   The period between the beginning of trading and the expiration of trading.

Limit   See *Position Limit; Price Limit; Reporting Limit; Variable Limit.*

Limit Move   A price that has advanced or declined the limit permitted during one trading session as fixed by the rules of a contract market.

Limit Order   An order in which the customer sets a limit on either price or time of execution, as contrasted with a market order, which implies that the order should be filled at the most favorable price as soon as possible.

Liquid Market   A market where selling and buying can be accomplished easily due to the presence of many interested buyers and sellers.

Liquidation   Offsetting positions by acquiring exactly opposite positions.

Liquidity (or Liquid Market)   A broadly traded market where buying and selling can be accomplished with small price changes and bid and offer price spreads are narrow.

Loan Program   The primary means of government agricultural price support operations, in which the government lends money to farmers at announced rates, with crops used as collateral. Default on these loans is the primary method by which the government acquires stocks of agricultural commodities.

Long   One who has bought a cash commodity or stock or a commodity futures contract or option, in contrast to a short, who has sold a cash entity, stock, futures, or option.

Long Hedge   Buying futures or option contracts to protect against possible increased prices of stocks or commodities. See also *Hedging*.

Maintenance Margin   The amount of money that must be maintained on deposit while a futures position is open. If the equity in a customer's account drops under the maintenance margin level, the broker must issue a call for money that will restore the customer's equity in the account to required initial levels. See also *Margin*.

Margin   In the futures industry, it is an amount of money deposited by both buyers and sellers of futures contracts to ensure performance against the contract. It is not a down payment.

Margin Call   A call from a brokerage firm to a customer to bring margin deposits back up to minimum levels required by exchange regulations; similarly, a request by the clearinghouse to a clearing member firm to make additional deposits to bring clearing margins back to minimum levels required by clearinghouse rules.

Mark-to-the-Market   Futures and options contracts in an account are extended daily using the settlement price, and the profit or loss is calculated.

Market Order   An order to buy or sell a stock, futures, or options contract which is to be filled at the best possible price and as soon as possible; in contrast to a limit order, which may specify requirements for price or time of execution. See also *Limit Order*.

Maturity   The period within which a futures contract can be settled by delivery of the actual commodity; the period between the first notice day and the last trading day of a commodity futures contract.

Maximum Price Fluctuation   See *Limit Move*.

Minimum Price Fluctuation   See *Point*.

Misrepresentation   An untrue or misleading statement concerning a material fact relied upon by a customer when making his or her decision about an investment.

Momentum Indicator   A line that is plotted to represent the difference between today's price and the price a fixed number of days ago. Momentum can be measured as the difference between today's price and the current value of a moving average. Often referred to as momentum oscillators.

Moving Average   A mathematical procedure to smooth or eliminate the fluctuations in data. Moving averages emphasize the direction of a trend, confirm trend reversals, and smooth out price and volume fluctuations or "noise" that can confuse interpretation of the market.

Naked Writing   Writing a call or a put on a stock or futures contract in which the writer has no opposite cash or futures market position. This is also known as uncovered writing.

National Association of Futures Trading Advisors (NAFTA)   The national trade association of commodity pool operators (CPOs), commodity trading advisors (CTAs), and related industry participants.

National Futures Association (NFA)   The industrywide self-regulatory organization of the futures industry.

National Securities Dealers Association   The industrywide self-regulatory organization of the securities industries.

Nearby Delivery (Month)   The futures contract closest to maturity.

Nearbys   The nearest delivery months of a futures and options contracts.

Net Asset Value   The value of each unit of a trading pool. Basically, a calculation of assets minus liabilities plus or minus the value of open positions (marked-to-the-market) divided by the number of units.

Net Performance   An increase or decrease in net asset value exclusive of additions, withdrawals, and redemptions.

Net Position The difference between the open long (buy) contracts and the open short (sell) contracts held by any one person in any one futures contract month or in all months combined.

Neutral Spread Another name for a delta-neutral spread. Spreads may also be lot-neutral, where the total number of long contracts and the total number of short contracts of the same type are approximately equal.

Nominal Price The declared price for a futures or options contract, sometimes used in place of a closing price when no recent trading has taken place in that particular delivery month; usually an average of the bid and asked prices.

Nondisclosure Failure to disclose a material fact needed by the customer to make a decision regarding an investment.

Normalizing An adjustment to data, such as a price series, to put the data within normal or a more standard range. A technique used to develop a trading system.

Not Held An order submitted to a broker, and over which the broker has discretion as to the prices and time at which the order will be executed, if at all.

Notice Day See *First Notice Day*.

Notice of Delivery See *Delivery Notice*.

Offer An indication of willingness to sell at a given price; opposite of bid.

Offset The liquidation of a position in the stock or futures markets by taking an equal, but opposite, position. All specifications must be the same.

Omnibus Account An account carried by one futures commission merchant with another in which the transactions of two or more persons are combined, rather than designated separately, and the identity of the individual accounts is not disclosed.

Open The period at the beginning of the trading session officially designated by the exchange during which all transactions are considered made "at the open."

Open Interest The total number of futures contracts of a given commodity which have not yet been offset by opposite futures transactions nor fulfilled by delivery of the actual commodity; the total number of open transactions, with each transaction having a buyer and a seller.

Open Outcry A method of public auction for making bids and offers in the trading pits or rings of exchanges.

Open Trade Equity The unrealized gain or loss on open positions.

Opening Range The range of closely related prices at which transactions took place at the opening of a market session; buying and selling orders at the opening might be filled at any point within such a range.

Opening Transaction A purchase or sale that establishes a new position.

Option Class All option contracts of the same type, covering the same underlying futures contract, commodity, or security.

Option Contract A unilateral contract that gives the buyer the right, but not the obligation, to buy or sell a specified quantity of stocks or futures contracts at a specific price within a specified period of time, regardless of the current market price.

The seller of the option has the obligation to sell the stock or futures contract or buy it from the option buyer at the exercise price, if the option is exercised. See also *Call (Option)*; *Put (Option)*.

Option Premium   The money, securities, or property the buyer pays to the writer (grantor) for granting an option contract.

Option Seller   See *Grantor*.

Order Execution   The handling of a customer order by a broker—includes receiving the order verbally or in writing from the customer, transmitting it to the trading floor of the exchange where the transaction takes place, and returning confirmation (fill price) of the completed order to the customer.

Orders   See *Market Order*; *Stop Order*.

Original Margin   The term applied to the initial deposit of margin money required of clearing member firms by clearinghouse rules; parallels the initial margin deposit required of customers.

Out-of-the-Money   A call option with a strike price higher or a put option with a strike price lower than the current market value of the underlying asset.

Out Trade   A trade made on an exchange which cannot be processed due to conflicting terms reported by the two parties involved in the trade.

Overbought   A technical opinion that the market price has risen too steeply and too fast in relation to underlying technical or fundamental factors.

Oversold   A technical opinion that the market price has declined too steeply and too fast in relation to underlying technical or fundamental factors.

P&S Statement   See *Purchase and Sale Statement*.

Par   A particular price, 100 percent of principal value.

Parity   A theoretically equal relationship between farm product prices and all other prices. In farm program legislation, parity is defined in such a manner that the purchasing power of a unit of an agricultural commodity is maintained at its level during an earlier historical base period.

Pit   A specially constructed arena on the trading floor of some exchanges where trading in stocks, futures, or options contracts is conducted by open outcry. On some exchanges, the term "ring" designates the trading area.

Point   The minimum fluctuation in futures prices or options premiums.

Point Balance   A statement prepared by futures commission merchants to show profit or loss on all open contracts by computing them to an official closing or settlement price.

Pool   Investments of more than one individual grouped together. See *Commodity Pool*.

Position   A market commitment.

Position Limit   The maximum number of futures contracts in certain regulated commodities that one can hold, according to the provisions of the CFTC. See *Reporting Limit*.

Position Trader   A trader who either buys or sells contracts and holds them for an extended period of time, as distinguished from the day trader, who will normally initiate and liquidate a position within a single trading session.

Premium   (1) The additional payment allowed by exchange regulations for delivery or higher-than-required standards or grades of a commodity against a futures contract. (2) In speaking of price relationships between different delivery months of a given commodity, one is said to be trading at a premium over another when its price is greater than that of the other. (3) Also can mean the amount paid a grantor or writer of an option by the buyer.

Price Limit   The maximum price advance or decline from the previous day's settlement price permitted for a futures or option contract in one trading session by the rules of the exchange.

Primary Markets   The principal market for the purchase and sale of a cash commodity.

Principal   Refers to a person who is a principal of a particular brokerage company or entity. (1) Any person including, but not limited to, a sole proprietor, a general partner, officer, or director, or a person occupying a similar status or performing similar functions, having the power, directly or indirectly, through agreement or otherwise, to exercise a controlling influence over the activities of the entity. (2) Any holder or any beneficial owner of 10 percent or more of the outstanding shares of any class of stock of the entity. (3) Any person who has contributed 10 percent or more of the capital of the entity.

Private Wires   Wires leased by various firms and news agencies for the transmission of information to branch offices and subscriber clients.

Proceeding Clerk   The member of the CFTC's staff in the Office of Proceedings who maintains the commission's reparations docket, assigns reparation cases to an appropriate CFTC official, and acts as custodian of the records of proceedings.

Producer   A person or entity that produces (grows, mines, etc.) a commodity.

Public Elevators   Grain storage facilities, licensed and regulated by state and federal agencies, in which space is rented out to whoever is willing to pay for it; some are also approved by the commodity exchanges for delivery of commodities against futures contracts.

Purchase and Sale Statement (P&S)   A statement sent by a commission house to a customer when a futures or options position has been liquidated or offset. The statement shows the number of contracts bought or sold, the gross profit or loss, the commission charges, and the net profit or loss on the transaction. Sometimes combined with a confirmation statement.

Purchase Price   The total actual cost paid by a person for entering into a commodity option transaction, including premium, commission, or any other direct or indirect charges.

Put (Option)   An option that gives the option buyer the right but not the obligation to sell the underlying asset at a particular price on or before a particular date.

Pyramiding   The use of profits on an existing position as margins to increase the size of the overall position, normally in successively smaller increments; for example, the use of profits on the purchase of five futures contracts as margin to purchase an additional four contracts, whose profits will in turn be used to margin an additional three contracts.

Quotation   The actual price or the bid or ask price of either cash commodities or futures, stocks, or options contract at a particular time; often called quote.

Rally   An upward movement of prices. See also *Recovery*.

Rally Top   The point where a rally stalls. A bull move will usually make several rally tops over its life.

Range   The difference between the high and low price of a commodity during a given period, usually a single trading session.

Ratio Spread   Any spread where the number of long market contracts and the number of short market contracts are unequal.

Ratio Vertical Spread   A spread where more contracts are sold than are purchased, with all contracts having the same underlying entity and expiration date. Ratio vertical spreads are usually delta-neutral.

Ratio Writing   A strategy used by option writers. It involves writing a covered call option as well as one or more uncovered call options. One of its objectives is to reduce some of the risk of writing uncovered call options, since the covered call does provide some degree of protection.

Reaction   A short-term countertrend movement of prices.

Recovery   An upward movement of prices following a decline.

Registered Commodity Representative (RCR)   See *Broker*; *Associated Person* (*AP*).

Registered Representative   A stockbroker. Stockbrokers are registered with the NASD and by the state(s) in which they conduct business.

Regulations (CFTC)   The regulations adopted and enforced by the federal overseer of the futures markets, the Commodity Futures Trading Commission, in order to administer the Commodity Exchange Act.

Reparations   Compensation payable to a wronged party in a futures or options transaction. The term is used in conjunction with the Commodity Futures Trading Commission's customer claims procedure to recover civil damages.

Reparations Award   The amount of monetary damages a respondent may be ordered to pay to a complainant.

Reporting Limit   The sizes of positions set by the exchange and/or by the CFTC at or above which commodity traders must make daily reports to the exchange and/or the CFTC about the size of the position by commodity, by delivery month, and according to the purpose of trading, i.e., speculative or hedging.

Resistance   The price level where a trend stalls. Opposite of a support level. Prices must build momentum to move through resistance.

Respondents   The individuals or firms against which a complaint is filed and a reparations award is sought.

Retender   The right of holders of futures contracts who have been tendered a delivery notice through the clearinghouse to offer the notice for sale on the open market, liquidating their obligation to take delivery under the contract; applicable only to certain commodities and only within a specified period of time.

Retracement   A price movement in the opposite direction of the prevailing trend. See *Correction*.

Reverse Conversion (Reversal)   A short underlying position, together with a long call and short put, where both options have the same exercise price and expiration date. A reverse conversion is a short underlying position offset by a long synthetic underlying position.

Rho   The sensitivity of an option's theoretical value to a change in interest rates.

Ring   A circular area on the trading floor of an exchange where traders and brokers stand while executing futures or options trades. Some exchanges use pits rather than rings.

Round Lot   A quantity of a commodity equal in size to the corresponding futures contract for the commodity, as distinguished from a job lot, which may be larger or smaller than the contract. In stocks, a 100-share trade.

Round Turn   The combination of an initiating purchase or sale of a futures contract and offsetting sale or purchase of an equal number of futures contracts to the same delivery month. Commission fees for commodity transactions cover the round turn.

Rules (NFA or NASD)   The standards and requirements to which participants who are required to be members of the National Futures Association or National Association of Security Dealers must subscribe and conform.

Sample Grade   In commodities, usually the lowest quality acceptable for delivery in satisfaction of futures contracts. See *Contract Grades*.

Scalper   A speculator on the trading floor of an exchange who buys and sells rapidly, with small profits or losses, holding positions for only a short time during a trading session. Typically, a scalper will stand ready to buy at a fraction below the last transaction price and to sell at a fraction above, thus creating market liquidity.

Security Deposit   See *Margin*.

Segregated Account   A special account used to hold and separate customer's assets from those of the broker or firm.

Selling Hedge   Selling futures or option contracts to protect against possible decreased prices of commodities or stocks which will be sold in the future. See *Hedging; Short Hedge*.

Serial Expiration   Options on the same futures contract or stock which expire in more than one month.

Series   All options of the same class having the same exercise price and expiration date.

Settlement Price   The closing price, or a price within the range of closing prices, that is used as the official price in determining net gains or losses at the close of each trading session.

Short One who is expecting prices to decline. For example, a trader who has sold a cash commodity or a commodity futures contract or bought a put, in contrast to a long, who has bought a cash commodity or a futures contract or a call.

Short Hedge Selling futures or options to protect against possible decreasing prices of commodities or stocks. See also *Hedging*.

Speculator One who attempts to anticipate price changes and make profits through the sale and/or purchase of stocks, commodity futures, or option contracts.

Spot Market for the immediate delivery of the product and immediate payment. May also refer to the nearest delivery month of a futures or options contract.

Spot Commodity See *Cash Commodity*.

Spread (or Straddle) The purchase of one futures or option delivery month against the sale of another. For example, the purchase of one delivery month of one option against the sale of the same delivery month of a different option. See also *Arbitrage*.

Stock-Type Settlement A settlement procedure in which the purchase of a contract requires immediate and full payment by the buyer to the seller. In a stock-type settlement the actual cash profit or loss from a trade is not realized until the position is liquidated.

Stop Loss A risk management technique used to close out a losing position at a given point. See *Stop Order*.

Stop Order An order that becomes a market order when a particular price level is reached. A sell stop is placed below the market; a buy stop is placed above the market. Sometimes referred to as a stop-loss order.

Straddle A trading strategy. A straddle involves writing a put as well as a call on the same stock or futures. Both options also carry the same strike price and the same expiration date.

Strike Price See *Exercise Price*.

Support A price level at which a declining market has stopped falling. Opposite of a resistance price range. Once this level is reached, the market usually trades sideways for a period of time.

Switch The liquidation of a position in one delivery month of a commodity and the simultaneous initiation of a similar position in another delivery month of the same commodity. When used by hedgers, this tactic is referred to as "rolling forward" the hedge.

Synthetic Call A long (short) underlying position together with a long (short) put.

Synthetic Put A short (long) underlying position together with a long (short) call.

Synthetic Underlying A long (short) call together with a short (long) put where both options have the same underlying exercise price and expiration date.

Technical Analysis An approach to analyze the stock and futures markets. Technicians normally examine such indicators as price charts patterns, rates of price changes, and changes in volume of trading and open interest.

Tender   The act on the part of the seller of futures contracts of giving notice to the clearinghouse that he or she intends to deliver the physical commodity in satisfaction of the futures contract. The clearinghouse in turn passes along the notice to the oldest buyer of record in that delivery month of the commodity. See also *Retender*.

Theoretical Value   An option value generated by a mathematical model given certain prior assumptions about the terms of the option, the characteristics of the underlying entity, and the prevailing interest rates. The most commonly used formula is known as the Black-Scholes.

Theta   The sensitivity of an option's theoretical value to a change in the amount of time to expiration.

Tick   Refers to a change in price up or down. See also *Point*.

Ticker Tape   A continuous paper-tape transmission of commodity or security prices, volume, and other trading and market information which operates on private or lease wires by the exchanges, available to their member firms and other interested parties on a subscription basis. Nowadays, it has been replaced by more efficient electronic transmission media.

Time Spread   The purchase and sale of options with the same exercise price on the same stock or futures contract, but with different expiration dates; also known as calendar and horizontal spreads.

Time Value   Any amount by which an option premium exceeds the option's intrinsic value; sometimes called extrinsic value or time premium.

To-Arrive Contract   A type of deferred shipment in which the price is based on delivery at the destination point and the seller pays the freight in shipping it to that point.

Traders   (1) People who trade for their own account. (2) Employees of dealers or institutions who trade for their employer's account.

Trading Range   An established set of price boundaries with a high price and a low price that a market will spend a marked period of time within.

Transferable Notice   See *Retender*.

Trend Line   A line drawn that connects a series of either highs or lows in a trend. The trend line can represent either support, as in an uptrend line, or resistance, as in a downtrend line. Consolidations are marked by horizontal trend lines.

Unauthorized Trading   The purchase or sale of stocks, commodity futures, or options for a customer's account without the customer's permission.

Underlying Futures Contract   The specific futures contract that the option conveys the right to buy (in the case of a call) or sell (in the case of a put).

Underlying Security   The security underlying an option. It would be purchased or sold were the option exercised.

Variable Limit   A price system that allows larger than normally allowed price movements under certain conditions. In periods of extreme volatility, some exchanges permit trading and price levels to exceed regular daily limits. At such times, margins may be automatically increased.

Variation Margin Call   A mid-season call by the clearinghouse on a clearing member requiring the deposit of additional funds to bring clearing margin monies up to minimum levels in relation to changing prices and the clearing member's net position.

Vega   The sensitivity of an option's theoretical value to a change in volatility.

Vertical Spread   A spread in which one option is bought and one option is sold, where both options are of the same type, have the same underlying entity, and expire at the same time. The options differ only by their exercise prices.

Volatility   A measure of the tendency of the price of an asset (stock, option, etc.) to move up and down, based on its daily price history over a period of time.

Volume of Trade   The number of contracts traded during a specified period of time.

Warehouse Receipt   A document guaranteeing the existence and availability of a given quantity and quality of a commodity in storage; commonly used as the instrument of transfer of ownership in both cash and futures transactions.

Wirehouse   See *Futures Commission Merchant (FCM)*.

Write or Writer   See *Grantor*.

Writing   The sale of an option in an opening transaction.

Note: This Glossary is included to assist the reader. It is not a set of legal definitions, nor a guide to interpreting the Commodity or Securities Exchange Act or any other legal instrument. For all legal assistance, contact your personal attorney.

APPENDIX 2

Sources of More Information

There is no shortage of material about options to study. Here is a brief reading list:

BOOKS

Babcock, Bruce, Jr., *The Business One Irwin Guide to Trading Systems*, Business One Irwin, Homewood, 1989.

Bernstein, Jake, *Facts on Futures: Insights and Strategies for Winning in the Futures Markets*, Probus Publishing Company, Chicago, 1987.

Bookstaber, Richard M., *Option Pricing and Investment Strategies*, 3rd ed., Probus Publishing Company, Chicago, 1991.

Caplan, David L., *The Options Advantage: Gaining a Trading Edge over the Markets*, Probus Publishing Company, Chicago, 1991.

Chance, Dom M., *An Introduction to Options and Futures*, Dryden Press, Chicago, 1989.

Derosa, David F., *Options on Foreign Exchange*, Probus Publishing Company, Chicago, 1992.

Elliot, R. N., *The Wave Principle*, Elliott, New York, 1938.

The Encyclopedia of Historical Charts, Commodity Perspective, Chicago.

Futures and Options Fact Book, Futures Industry Institute, Washington, DC, 1992.

Gann, William D., *How to Make Profits in Commodities*, Lambert-Gann, Pomeroy, WA, 1951.

Gann, William D., *Truth of the Stock Tape*, Financial Guardian, New York, 1932.

Labuszewski, John W., and Sinquefield, Jeanne Cairns, *Inside the Commodity Option Markets*, John Wiley & Sons, New York, 1985.

Mehrabian, Albert, *Your Inner Path to Investment Success: Insights into the Psychology of Investing*, Probus Publishing Company, Chicago, 1991.

Natenberg, Sheldon, *Option Volatility and Pricing Strategies: Advanced Trading Techniques for Professionals*, Probus Publishing Company, Chicago, 1988.

Petzel, Todd E., *Financial Futures and Options: A Guide to Markets, Applications, and Strategies*, Quorum Books, New York, 1989.

Schwager, Jack D., *A Complete Guide to the Futures Markets: Fundamental Analysis, Technical Analysis, Trading, Spreads, and Options*, John Wiley & Sons, New York, 1984.

Schwager, Jack D., *Market Wizards: Interviews with Top Traders*, Simon & Schuster, New York, 1989.

Schwager, Jack D., *The New Market Wizards: Conversations with America's Top Traders*, Harper Business, New York, 1992.

NEWSLETTERS AND MAGAZINES

Cycles, *Foundation for the Study of Cycles,* 2600 Michelson Drive, Suite 1570, Irvine, CA 92715.

Opportunities in Options Newsletter, P.O. Box 2126, Malibu, CA 90265.

Technical Analysis of Stocks and Commodities, 9131 California Avenue SW, Seattle, WA 98136.

AGENCIES

Alliance Against Fraud Telemarketing
c/o National Consumers League
815 15th Street NW
Suite 516
Washington, DC 20005

American Association of Individual Investors
612 North Michigan Avenue
Chicago, IL 60611
312/280–0170

Commodity Futures Trading Commission
2033 K Street NW
Washington, DC 20581
202/254–6387

Council of Better Business Bureaus
1515 Wilson Boulevard
Arlington, VA 22209
703/276–0100

Federal Bureau of Investigation
Justice Department
9th Street and Pennsylvania Avenue NW
Washington, DC 20580
202/324–3000

Federal Trade Commission
6th Street and Pennsylvania Avenue NW
Washington, DC 20580
202/326–2222

National Association of Securities Dealers
1735 K Street NW
Washington, DC 20006
202/728–8044

National Consumers League
815 15th Street NW
Suite 516
Washington, DC 20006
202/639–8140

National Futures Association
200 W. Madison
Suite 1600
Chicago, IL 60606
Toll-free: 800/621–3570
In IL: 800/621–3570

National Association of Securities Dealers, Inc.
1735 K Street NW
Washington, DC 20006–1506
202/728–8000

North American Securities Administration Association
2930 SW Wanamaker Drive
Suite 5
Topeka, KS 66614
913/273–2600

The Options Industry Council
440 South LaSalle Street
Suite 2400
Chicago, IL 60605

Securities and Exchange Commission
450 Fifth Street NW
Washington, DC 20006
202/728–8233

United States Postal Service
Chief Postal Inspector
Room 3021
Washington, DC 20260–2100
202/268–4267

TRADING SCHOOLS (ALSO CONTACT EXCHANGES)

International Trading Institute
440 South LaSalle Street
Suite 3112
Chicago, IL 60605
312/663–8999

Exchanges and Options Offered

All exchanges have public information, public relations, marketing, or sales departments, which prepare and distribute a wide variety of instructional material. Most of them conduct seminars to teach investors how to trade the options listed on their exchange. They are an excellent source to obtain additional information.

STOCK OPTIONS

Since there are so many options offered, you need to contact one of the exchanges below for the 76-page *Directory of Exchange Listed Options.*

AMERICAN STOCK EXCHANGE
Derivative Securities
86 Trinity Place
New York, NY 10006
1–800-THE-AMEX

or

2, London Wall Buildings
London Wall
London EC2M 5SY ENGLAND
44–71–628–5982

CHICAGO BOARD OPTIONS EXCHANGE
LaSalle at Van Buren
Chicago, IL 60605
1–800–OPTIONS
(312) 786–5600

NEW YORK STOCK EXCHANGE
Options and Index Products
11 Wall Street
New York, NY 10005
1–800–692–6973
(212) 656–8533

THE OPTIONS CLEARING CORPORATION
440 South LaSalle Street
Suite 2400
Chicago, IL 60605
1–800–537–4258
(312) 322–6200

PACIFIC STOCK EXCHANGE
Options Marketing
115 Sansome Street, 7th Floor
San Francisco, CA 94104
1–800-TALK-PSE
(415) 393–4028

PHILADELPHIA STOCK EXCHANGE
Philadelphia Board of Trade
(Foreign Currency Options)
1900 Market Street
Philadelphia, PA 19103
1–800-THE-PHLX
(215) 496–5404

or

PHLX European Office
39 King Street
London EC2V 8DQ ENGLAND
44–71–606–2348

OPTIONS-ON-FUTURES EXCHANGES

CHICAGO BOARD OF TRADE
141 West Jackson Boulevard
Chicago, IL 60604
(312) 435–3500

CHICAGO MERCANTILE EXCHANGE
Index & Option Market
International Monetary Market
30 South Wacker Drive
Chicago, IL 60606
(312) 930–8200

COFFEE, SUGAR & COCOA EXCHANGE, INC.
Four World Trade Center
New York, NY 10048
(212) 938–2800

COMMODITY EXCHANGE INC.
Four World Trade Center
New York, NY 10048
(212) 938–2900

KANSAS CITY BOARD OF TRADE
4800 Main Street
Kansas City, MO 64112
(816) 753–7500

MIDAMERICA COMMODITY EXCHANGE
141 West Jackson Boulevard
Chicago, IL 60604
(312) 341–3000

MINNEAPOLIS GRAIN EXCHANGE
150 Grain Exchange Building
Minneapolis, MN 55415
(612) 338–6212

NEW YORK COTTON EXCHANGE
Four World Trade Center
New York, NY 10048
(212) 938–2650

NEW YORK FUTURES EXCHANGE
30 Broad Street
New York, NY 10004
(212) 623–4949

NEW YORK MERCANTILE EXCHANGE
Four World Trade Center
New York, NY 10048
(212) 938–2222

Specifications for Exchange Traded Option-on-Futures for U.S. Markets

COMMODITY	EXCH.	TRADING HRS Central time	DELIVERY MONTHS	CONTRACT SIZE	TICK SIZE	DAILY LIMITS	STRIKE PRICE INCREMENTS
Australian Dollar	IMM	7:20-2:00	F, H, J, M, N, U, V, Z	100,000 AD	1pt = $10.00	150 pts	Every 50 pts (i.e., 7350, 7400, 7450)
British Pound	IMM	7:20-2:00	H, M, U, Z	62.50 BP	02¢/£ = $12.50	400 pts	Every 02¢/£ (i.e., 170, 172, 174)
Canadian Dollar	IMM	7:20-2:00	H, M, U, Z	100,000 CD	1pt = $10.00	100 pts	Every ½¢ (i.e., 7300, 7350)
Cattle, Fdr	CME	8:45-1:00	F, H, J, K, Q, U, V, X	50,000 lbs	2 1/2¢/cwt = $12.50	None	2¢ increments (i.e., 64¢, 66¢, 68¢)
Cattle, Live	CME	8:45-1:00	G, J, M, Q, V, Z	40,000 lbs	2 1/2¢/cwt = $10.00	None	2¢ increments (i.e., 64¢, 66¢, 68¢) back months
Cocoa (Metric)	NYCSCE	8:30-1:10	H, K, N, U, Z	10M ton (22,046)	1pt = $10.00	None	$50 increments for prices below $1500 / $100 increments for prices between $1500 & $3600 / $200 increments for prices above $3600
Coffee	NYCSCE	8:15-12:58	H, K, N, U, Z	37,500 lbs	1¢/lb = $375	None	025 increments for prices below $1.00 / 05 increments for prices between $1.00 and $2.00 / 10 increments for prices above $2.00
Copper, High Grade	COMEX	8:25-1:00	F, H, K, N, U, Z	25,000 lbs	5pt = $12.50	None	1¢/lb increments for prices below 40¢ / 2¢/lb increments for prices between 40¢ and $1.00 / 5¢/lb increments for prices above $1.00
Corn (2)	CBOT	8:30-1:45	H, K, N, U, Z	5,000 bu	1/8¢ = $6.25	10¢ = $500(4)	Every 10¢ per bushel (i.e., $2.80, $2.90, $3.00)
Corn	MA	8:30-1:45	H, K, N, U, Z	1,000 bu	1/8¢ = $1'.25	10¢ = $100(4)	Every 10¢ per bushel (i.e., $2.80, $2.90, $3.00)
Cotton	NYCTE	9:30-1:40	H, K, N, V, Z	50,000 lbs	1¢/cwl = $5.00	None	1¢ increments for prices 74¢ and below / 2¢ increments for prices 76¢ and above
CRB Index	NYFE	8:45-1:45	H, K, N, U, Z	500xCRB Index	5pts = $25.00	None	Less than $300.00—$5.00 increments / Above $300.00—$10.00 increments
Crude Oil	NYME	8:45-2:10	All Months	1,000 barrels	1pt = $10.00	None	100 point intervals
Deutschemark	IMM	7:20-2:00	H, M, U, Z	125,000 DM	1pt = $12.50	150 pts	Every 50 points (i.e., 5500, 5550, 5600)
Eurodollar	IMM	7:20-2:00	H, M, U, Z	$1,000,000	01 index pt = $25.00	None	25 point increments
Gasoline, Unleaded	NYME	8:50-2:10	All Months	42,000 gal	1pt = $4.20	None	200 point intervals
Gold	COMEX	7:20-1:30	G, J, M, Q, V, Z	100 Troy oz	10¢/oz = $10.00 (option quote price in $/oz)	None	$10 increments for prices below $500 / $20 increments for prices between $501 and $1000 / $50 increments for prices above $1000
Gold	MA	7:20-1:40	F, H, J, M, N, U, V, Z	33.2 troy oz	10¢/oz = $3.32	None	$10 increments for prices below $400 / $20 increments for prices between $400 and $600 / $30 increments for prices between $600 and $900 / $40 increments for prices above $900
Heating Oil	NYME	8:50-2:10	All Months	42,000 gal	1pt = $4.20	None	200 point intervals
Hogs	CME	8:45-1:00	G, J, M, N, Q, V, Z	40,000 lbs	2.5¢ cwl = $10.00	None	2¢ increments (i.e., 48¢, 50¢, 52¢)
Japanese Yen (1)	IMM	7:20-2:00	H, M, U, Z	12,500,000JY	1pt = $12.50	150 pts	Every 50 points (i.e., 7400, 7450, 7500)
Lumber	CME	9:00-1:05	F, H, K, N, U, Z	160,000 bd ft	10¢ = $1.60	None	$5.00 increments
MMI Index	CBOT	8:15-3:15	All Months	500 x MMI	5pts = $25.00	15 index points	5 index points
Municpal Bonds (2)	CBOT	7:20-2:00	H, M, U, Z	$1000 x BB Index	1/64 = $15.63	96/32 = $3000	1 point intervals (i.e., 8800, 8900, 9000)
Nikkie Index	CME	8:00-3:15	H, M, U, Z	$5.00 x Index	5pts = $25.00	None	Every 500 pts (i.e., 275.00, 280.00, 285.00)
NYSE Index	NYFE	8:30-3:15	H, M, U, Z	500 x Index	5pts = $25.00	None	Every 200 points (i.e., 94, 96, 98)

Commodity	Exchange	Hours	Months	Size	Tick	Limit	Increments
Oats (2)	CBOT	9:30–1:15	H, K, N, U, Z	5,000 bu	1/8¢ = $6.25	10¢ = $500	Every 10¢ per bushel
Oats	MA	9:30–1:45	H, K, N, U, Z	1,000 bu	1/8¢ $1.25	10¢ = $100	Every 10¢ per bushel (i.e., 200, 230, 240)
Orange Juice	NYCTE	9:15–1:15	F, H, K, N, U, X	15,000 lbs	5pts = $7.50	None	Every 250 pts (i.e., 122.50, 125.00, 127.50)
Platinum	NYME	7:20–1:30	F, J, N, V	50 troy oz	$0.10 = $5.00	None	$20.00 per troy oz
Pork Bellies	CME	8:45–1:00	G, H, K, N, Q	40,000 lbs	.00025¢/lb = $10.00	None	2¢ per pound intervals (i.e., 64¢, 66¢, 68¢)
S&P 500 Index	IOM	8:30–3:15	H, M, U, Z	500xS&P Index	5pts = $25.00	None	Every 500 points (1000 pts in 3rd and 4th nearest month in quarterly cycle.)
Silver	COMEX	7:25–1:25	F, H, K, N, U, Z	5,000 troy oz	.001¢/oz = $25.00	None	20¢ increments for prices below $8.00 / 50¢ increments for prices between $8.00 and $15.00 / $1.00 increments for prices $15.00 and above
Silver, New (2)	CBOT	7:25–1:25	G, J, M, Q, V, Z	1,000 troy oz	.01¢/oz = $1.00	$1.00 = $1000(3)	25¢/oz increments for prices below $8 / 50¢/oz increments for prices from $8 to $20 / $1.00/oz increments for prices above $20
Soybean Meal	CBOT	8:30–1:15	F, H, K, N, Q, U, V, Z	100 tons	5¢/ton = $5.00	$10/1000pts(4)	$5 if less than $200 $10 if more than $200
Soybean Oil (2)	CBOT	8:30–1:15	F, H, K, N, Q, U, V, Z	60,000 lbs	1pt = $6.00	1¢/100pts(4)	Less than 2500 pts 50 pt increments / More than 2500 pts 100 pt increments
Soybeans	CBOT	8:30–1:15	F, H, K, N, Q, U, X	5,000 bu	1/8¢ = $6.25	30¢ = $1500(4)	Every 25¢ per bushel (i.e., $7, $7.25, $7.50)
Soybeans	MA	8:30–1:15	F, H, K, N, Q, U, X	1,000 bu	1/8¢ = $1.25	30¢ = $300	Every 25¢ per bushel (i.e., $7, $7.25, $7.50)
Sugar	NYCSCE	9:00–12:50	H, K, N, U, V	112,000 lbs	.01¢ = $11.20	None	1/2¢ increments for prices below 10¢ / 1¢ increments for prices between 11¢ and 40¢ / 2¢ increments for prices 42¢ and over
Swiss Franc (1)	IMM	7:20–2:00	H, M, U, Z	125,000 SF	.01¢/SF = $12.50	150 pts	Every 50 points (i.e., 7750, 7800, 7850)
T-Bills 90-Days (1)	CME	7:20–2:00	H, M, U, Z	$1,000,000	1pt = $25.00	None	25 point increments
T-Notes 5/yr	CBOT	7:20–2:00	H, M, U, Z	$100,000	1/64¢ = $15.63	96/32 = $3000(4)	1/2¢ point ($500) (i.e., 93.00, 93.50, 94.00)
Treasury Bonds	MA	7:20–3:15	H, M, U, Z	$50,000	1/64¢ = $7.81	96/32 = $1500	Every 2 full points (i.e., 64, 66,68)
Treasury Notes	CBOT	7:20am–2:00pm 6:00pm–9:30pm	H, M, U, Z	$100,000	1/64¢ = $15.63	96/32 = $3000 (4)	Every 1 full point (i.e., 77, 78, 79)
U.S. Dollar Index	NYCTE	7:20–2:00	H, M, U, Z	500 x Index	1pt = $10.00	None	Every 2¢ (i.e, 100, 102, 104)
Wheat	CBOT	8:30–1:15	H, K, N, U, Z	5,000 bu	1/8¢ = $6.25	20¢ = $1000(4)	Every 10¢ per bushel (i.e., $3.50, $3.60, $3.70)
Wheat	KCBT	8:30–1:15	H, K, N, U, Z	5,000 bu	1/8¢ = $6.25	25¢ = $1250	Every 10¢ per bushel (i.e., $3.50, $3.60, $3.70)
Wheat	MPLS	8:30–1:45	H, K, N, U, Z	5,000 bu	1/8¢ = $6.25	25¢ = $1250	Every 10¢ per bushel (i.e., $3.50, $3.60, $3.70)
Wheat	MA	8:30–1:45	H, K, N, U, Z	1,000 bu	1/8¢ = $1.25	20¢ = $200	Every 10¢ per bushel (i.e., $3.50, $3.60, $3.70)

Exchanges

CBOT = Chicago Board of Trade
CME = Chicago Mercantile Exchange
COMEX = Commodity Exchange, Inc (NY)
IMM = International Monetary Market, Div. of CME
IOM-Index and Options Market, Div. of CME
KCBT = Kansas City Board of Trade
MA = Mid America Commodity Exchange
MPLS = Minneapolis Board of Trade
NYCSCE = New York Coffee, Sugar and Cocoa Exchange
NYCTE = New York Cotton Exchange
NYFE = New York Futures Exchange
NYME = New York Mercantile Exchange

Footnotes:
(1) Trades on GLOBEX. Electronic Trading Hours, CBOT Options 10:30 pm–6:00 am, CME 6:00 pm–6:00 am (CST)
(2) On last trading day, all CBOT options expire at noon (CST)
(3) No limit 2 days prior to delivery month
(4) No limit last trading day

Abbreviations for Months

Month	Current Year	Following Year
JAN	F	D
FEB	G	E
MAR	H	I
APR	J	L
MAY	K	O
JUN	M	P
JUL	N	T
AUG	Q	R
SEP	U	B
OCT	V	C
NOV	X	W
DEC	Z	Y

The information contained in this table is from sources believed to be reliable, but we cannot be held responsible for either its accuracy or completeness. All contact specifications are subject to change by action of the respective commodity exchange. Therefore, the specifications can be changed, sometimes without notice. You need to be in touch with your broker before entering any trade. Limits are subject to change and to variable limit rules of the respective exchange.

Description of Option Trading Software

Editor's Note: Prices are always subject to change without notice.

TRAINING PROGRAMS (MARKET SIMULATORS)

SOFTSware
Coffee, Sugar & Cocoa Exchange, Inc.
Four World Trade Center
New York, NY 10048
1–800-HEDGE IT
(212) 938–2966
Price: Free

Here's an excellent tutorial to bring you up to speed on options quickly. In one mode, it suggests the best strategies based on given market conditions. In the next mode, it lets you build the strategy you think will be most profitable. In both situations you see the profit and loss results at expiration for a range of prices. The graphs are particularly useful in helping new traders visualize the P/L relationships. Includes contract specifications for coffee, sugar, and cocoa options.

"The Game"
COMEX
Commodity Exchange, Inc.
Publications Department
Four World Trade Center
New York, NY 10048
1–800-THE-NYFE
(212) 938–2900
Price: $69

This program simulates the trading of gold and silver options on COMEX. You trade $10,000 of equity as if it were your own, selecting strategies, deciding on quantity, entering and exiting the markets. Find out if you have what it takes—without risking real money in the markets!

TRADING PROGRAMS

AIQ Incorporated
P.O. Drawer 7530
Incline Village, NV 89452–9934
1–800–332–2999
Price: $898

AIQ (Stock) Option Expert analyzes daily stock prices and volume data for all the stocks in your database and recommends which ones to consider for trading. The system then evaluates the options for each stock you have selected and recommends the best strategy or positions to take. It also recommends when to offset positions. Being an expert system, it calculates more than 30 technical indicators and then evaluates this price analysis against 160 rules to generate buy-sell recommendations. Needless to say, AIQ is a very sophisticated system. Additionally, it can identify leading/lagging groups of stocks, equity, and index option opportunities. The Profit Manager module is designed to protect your principal and profits. The graphics are extensive and impressive.

COMPUTRAC
CompuTrac Software, Inc.
1017 Pleasant Street
New Orleans, LA 70115
800–274–4028
(504) 895–1474
Price: $695 to $1,895

"The first is still the best" is their slogan. This software takes a very large toolbox approach to technical trading. It provides just about all the major studies with excellent documentation. It's for the very serious technician. The list of studies available can be found in Chapter 8.

Harloff Inc.
26106 Tallwood Drive
North Olmsted, OH 44070
(216) 734–7271
Price: $199

Call/put options software utilizes a proprietary formula for option evaluation of American-style options. The valuations are valid at any time in the life of the option, including expiration. It computes hedge ratios and breakeven hedge stock prices. It will also determine optimum times to buy and sell puts and calls. A stock option product.

Options–80
15 Stow Street
P.O. Box 471
Concord, MA 01742
(508) 369–1589
Price: $150

Options–80 is for listed stock options. The program analyzes calls, puts, and spreads; does Black-Scholes modeling; and calculates market-implied volatility. It plots the annualized return on investment against the expiration price of the underlying stock, guiding the user to optimum investment. Unique algorithms account for future payments, as well as buying and selling costs and the time value of money. Presents tables and charts for choosing transactions to give the highest yield for price action that the user thinks most likely. It is good for investigating "what-if" scenarios. Prints out, stores to disk, and can be backed up. Comes with a comprehensive indexed manual.

International Advanced Modes Inc.
P.O. Box 1019
Oak Brook, IL 60522
(708) 369–8461
Price: $124

OptionExpert—The strategist analyzes stocks and stock index option. Utilizes the Black-Scholes model. Contains provisions for taking commissions and dividends into account. On-line data retrieval is not required. Calculates deltas and volatilities, which can be adjusted and analyzed. Calculates and ranks strategies, based on risk and potential reward. Program is menu-driven, requiring no memorization of commands.

Optionomics Corporation
2835 East 3300 South
Suite 200
Salt Lake City, UT 84109
(801) 466–2111
(800) 255–3374
Price: $350–$995 per month

Optionomics Systems, which includes Portfoliomic and Strategist, analyzes options in real time, alerting traders to the impact of changes in implied volatility, deltas, gammas, vegas, and thetas. It also generates and evaluates strategies and allows the operator to ask a wide variety of what-if questions. Portfoliomics manages, evaluates, and tracks diversified option portfolios. Very strong in risk management.

Montgomery Investment Group
P.O. Box 508
Wayne, PA 19087–0508
(215) 688–2508
Price: $395 to $995

If you are a spreadsheet analyst, you'll love Montgomery's product line. The products are add-ons to Excel and Lotus 1–2–3, not templates. Functions are added to these spreadsheet programs to calculate fair-market value using seven of the most commonly used formulas, plus implied and historical volatility, deltas, gammas, thetas, and vegas. Evaluates options on stocks, bonds, foreign exchanges, commodities, futures, indexes, derivatives, real estate, scores, and warrants. Additionally, it can link to a variety of price quotation services for updates.

Discount Corporation of New York Futures
Madison Plaza
200 West Madison, 2nd Floor
Chicago, IL 60606
(312) 368–2200
Price: $1,495

Opposition Position Manager tests strategies, and evaluates portfolios under various scenarios by changing futures prices, volatility, and time to expiration. Displays risk-reward potential in tabular form, two- and three-dimensional graphs on screen or paper. Calculates theoretical prices, deltas, gammas, zetas (vegas), and thetas over a range of prices and time intervals. Uses either average volatility of markets or the individual option. Tracks open trade quantity, average trade price, settlement price, implied volatility for each option, unrealized gains (losses), and commissions.

OPA Software
P.O. Box 90658
Los Angeles, CA 90009
(213) 545–3716
(800) 321–4100
Price: $195–$750

Option Pricing Software is a reasonably priced package that claims the ability (expert system) of not only calculating fair-market value of stocks, indexes, and futures option contracts, but selecting the best option and strategy. Additionally, OPS analyzes straddles, combinations, and spreads. It generates maximum profits-losses, margin requirements, deltas, and omegas. No limitation is placed on the type of strategy to be studied. It's set up to run repeated what-if scenarios.

OptionVue Systems International, Inc.
175 East Hawthorn Parkway, Suite 180
Vernon Hills, IL 60061
(708) 816–6610
(800) 733–6610

OptionVue IV software is designed for the sophisticated stock option trader interested in stocks, stock indexes, cash-carry commodities, convertible securities, warrants, and arbitrage. The program utilizes a dividend-adjusted Black-Scholes formula. The uniqueness of the program is its ability to calculate what-if runs on its own. This function generates "all" the possible opportunities (outright positions, spreads, etc.) and ranks them as to potential return and risk. Includes a communications module to update price information automatically and 1½ years of volatility history on each optioned asset in the United States, so traders can study volatility patterns. An add-on module accommodates futures options and real-time price quotations.

Terco Computer Systems
P.O. Box 1803
Lombard, IL 60148
(708) 347–9182
Price: $95

OptionMaster calculates the fair-market premium and deltas for stocks, futures, bonds, and physical commodities for put and call contracts. Uses Black-Scholes and Cox-Ross-Rubenstein (binomial) formulas. Ideal for determining implied volatility or fair-market value arrays to spot overvalued or undervalued options.

Revenge Software, Inc.
P.O. Box 1073
Huntington, NY 11743
Price: $145

Option Valuator—Option Writer software has been written for the stock option trader. It provides the capability of calculating the information needed to evaluate options, primarily option value and volatility. The Option Writer part of the program enables the investor to evaluate and track a covered stock option writing program.

$ Dollar/Soft
P.O. Box 822
Newark, CA 94560–0822
(415) 656–5649
(800) 678–4664
Price: $99–$199

Option Trader classifies itself as "an expert system for the put and call options trader." For pricing of calls, it utilizes the Black-Scholes formula for stock and indexes. For calls in the futures markets, it uses the Fisher Black formula. Puts are calculated by the Academic or Conversion models. The program handles a variety of spreads (bull/bear, butterfly, ratio, horizontal, diagonal, etc.), as well as net long-short strategies. Includes calculations for writer's margin and return.

Institute for Options Research, Inc.
P.O. Box 6586
Lake Tahoe, NV 89449
(800) 334–0854, Ext 840
(702) 588–3590
Price: $89.95

Option Master is an affordable program for the novice option trader. This program allows you to calculate the volatility of the stock, futures, or index options you are considering. Once you do this, you can calculate the fair market value (FMV) for one or all of the options' strike prices. By comparing the FMV with the ask price, you locate options that appear to be overpriced or underpriced. The institute sells a complete line of books, audiotapes and videotapes, and software programs. Call for a catalog.

Townsend Analytics, Ltd.
100 South Wacker Drive, Suite 1506
Chicago, IL 60606
(312) 621–0141
(800) 827–0141
Price: $2,000

Option Risk Management evaluates risk to return of option portfolios of large traders (stocks, futures, stock indexes, debt instruments). Risk and return are measured by thetas, vegas, gammas, deltas, option valuations, and profit-loss. Requires Microsoft Windows 3.0. Townsend also sells a data service to link Option Risk Management to live quotation services.

Thomas A. McCafferty's involvement in the cash commodities, futures, and securities industries goes back to 1973. He has traded stocks, futures, and options for his own account and for others and has supervised brokers who traded for the public. Additionally, he has a strong background in sales and marketing and is the author of *In-House Telemarketing: A Master Plan for Starting and Managing a Profitable Telemarketing Program* (2nd edition, 1992). Other financial books written or coauthored by him include *Winning with Managed Futures, All about Futures,* and *All about Commodities.* Mr. McCafferty, as a branch office manager for Securities Corporation of Iowa, was registered as a futures broker, a securities broker, a securities and option principal, and an insurance broker. He has also been a real estate broker, a member of the marketing faculty of Upper Iowa University, and a consultant for the Small Business Administration. He lives and writes in Denver, Colorado. Mr. McCafferty is currently working on a book and a newsletter on scale trading selected commodity markets.